Research Issues in
Undergraduate Mathematics Learning

Preliminary Analyses and Results

Research Issues in Undergraduate Mathematics Learning

Preliminary Analyses and Results

James J. Kaput
and
Ed Dubinsky,
Editors

MAA Notes Number 33

Published and Distributed by
The Mathematical Association of America

MAA Notes and Reports Series

The MAA Notes and Reports Series, started in 1982, addresses a broad range of topics and themes of interest to all who are involved with undergraduate mathematics. The volumes in this series are readable, informative, and useful, and help the mathematical community keep up with developments of importance to mathematics.

MAA Notes

1. Problem Solving in the Mathematics Curriculum, *Committee on the Teaching of Undergraduate Mathematics,* a subcommittee of the Committee on the Undergraduate Program in Mathematics, *Alan H. Schoenfeld,* Editor
2. Recommendations on the Mathematical Preparation of Teachers, *Committee on the Undergraduate Program in Mathematics, Panel on Teacher Training.*
3. Undergraduate Mathematics Education in the People's Republic of China, *Lynn A. Steen,* Editor.
5. American Perspectives on the Fifth International Congress on Mathematical Education, *Warren Page,* Editor.
6. Toward a Lean and Lively Calculus, *Ronald G. Douglas,* Editor.
8. Calculus for a New Century, *Lynn A. Steen,* Editor.
9. Computers and Mathematics: The Use of Computers in Undergraduate Instruction, *Committee on Computers in Mathematics Education, D. A. Smith, G. J. Porter, L. C. Leinbach, and R. H. Wenger,* Editors.
10. Guidelines for the Continuing Mathematical Education of Teachers, *Committee on the Mathematical Education of Teachers.*
11. Keys to Improved Instruction by Teaching Assistants and Part-Time Instructors, *Committee on Teaching Assistants and Part-Time Instructors, Bettye Anne Case,* Editor.
13. Reshaping College Mathematics, *Committee on the Undergraduate Program in Mathematics, Lynn A. Steen,* Editor.
14. Mathematical Writing, by *Donald E. Knuth, Tracy Larrabee, and Paul M. Roberts.*
15. Discrete Mathematics in the First Two Years, *Anthony Ralston,* Editor.
16. Using Writing to Teach Mathematics, *Andrew Sterrett,* Editor.
17. Priming the Calculus Pump: Innovations and Resources, *Committee on Calculus Reform and the First Two Years,* a subcomittee of the Committee on the Undergraduate Program in Mathematics, *Thomas W. Tucker,* Editor.
18. Models for Undergraduate Research in Mathematics, *Lester Senechal,* Editor.
19. Visualization in Teaching and Learning Mathematics, *Committee on Computers in Mathematics Education, Steve Cunningham and Walter S. Zimmermann,* Editors.
20. The Laboratory Approach to Teaching Calculus, *L. Carl Leinbach et al.,* Editors.
21. Perspectives on Contemporary Statistics, *David C. Hoaglin and David S. Moore,* Editors.
22. Heeding the Call for Change: Suggestions for Curricular Action, *Lynn A. Steen,* Editor.
23. Statistical Abstract of Undergraduate Programs in the Mathematical Sciences and Computer Science in the United States: 1990–91 CBMS Survey, *Donald J. Albers, Don O. Loftsgaarden, Donald C. Rung, and Ann E. Watkins.*
24. Symbolic Computation in Undergraduate Mathematics Education, *Zaven A. Karian,* Editor.
25. The Concept of Function: Aspects of Epistemology and Pedagogy, *Guershon Harel and Ed Dubinsky,* Editors.
26. Statistics for the Twenty-First Century, *Florence and Sheldon Gordon,* Editors.
27. Resources for Calculus Collection, Volume 1: Learning by Discovery: A Lab Manual for Calculus, *Anita E. Solow,* Editor.
28. Resources for Calculus Collection, Volume 2: Calculus Problems for a New Century, *Robert Fraga,* Editor.

MAA Reports

These volumes may be ordered from:
The Mathematical Association of America
1529 Eighteenth Street, NW
Washington, DC 20036
800-331-1MAA   FAX 202-265-2384

©1994 by the Mathematical Association of America
ISBN 0-88385-090-7
Library of Congress Catalog Number 93-81062
Printed in the United States of America
Current Printing
10 9 8 7 6 5 4 3 2 1

INTRODUCTION

Mathematics education is in deep distress at all levels. Clarion calls for reform have pierced the consciousness of all who work in the mathematical sciences. Evidence of the need for change confronts us daily. Implementing reform — indeed, shaping that reform — demands the best understanding we can muster — understanding of student and teacher, of learning and thinking, of appropriate forms of instruction and instructional materials, of the processes of assessment, of appropriate institutional arrangements. The list of areas in which more knowledge — *research–based knowledge* — is desperately needed grows even as our knowledge grows.

This is why research is so important and why we should be aware of the kinds of contributions that research in undergraduate mathematics education can make. Through research, we can come to recognize systemic relationships that are not apparent from the single perspective of teaching practice. Research will help us to ask ever better questions and, at its best, provide important clues pointing towards possible answers. It is only through research into how people learn, and by comparing what happens with what our theoretical perspectives propose, that we can evaluate the outcomes of educational experiences for our students. A profound synthesis of research and practice is the real mark of progress in an intellectual discipline.

We are excited to be a part of the very beginning of development of such a synthesis, and one of our goals for this volume is to suggest to readers the possibility of getting in on the ground floor of a very good thing.

Research in Undergraduate Mathematics Education is a new and rapidly developing field. It is, therefore, without established structures. It needs to build conceptual frameworks and a supporting infrastructure of scholarly organizations, journals, graduate programs, as well as more informal networks of active scholars. This is a time for special sessions at large conferences, for informal get–togethers, and for ad hoc collections of papers. When a field is in transition, it is both difficult and necessary to try to situate research efforts in one or another global trend. This is a time for establishing, no matter how tentatively, conceptual frameworks and organizing principles that can help make sense of a growing number of reports and help guide future work.

This is how we viewed the situation in February 1990 when we asked the American Mathematical Society (AMS) and the Mathematical Association of America (MAA) to schedule a Joint Special Session on Research in Undergraduate Mathematics Education at their annual Winter Meeting to be held in San Francisco in January 1991. We felt that such a session would provide an opportunity for workers in this field to meet and interact in the standard scholarly manner. We also felt that it would be an opportunity to gain the attention of mainstream mathematicians. In both endeavors we were, at least initially, successful. The Special Session was held in a climate of dramatically increasing attention to matters educational. One of the concrete consequences of the Special Session is the existence of this book.

Through a combination of invitations and unsolicited submissions, we obtained about 40 presentations at the Session. We invited about 15 of the presenters to write up their reports as research papers and submit them for inclusion in this collection. We selected 9 of the submitted papers. We tried to choose those papers that contained the most important results and would be the most interesting for a diverse audience comprising of mathematicians, mathematics educators, and researchers in mathematics education. It was only after we completed our selection, based essentially on quality, that we tried to find some unifying themes that ran through the papers. To our delight, we found several.

TRANSITION THEME

The most important theme that we found has to do with the fact that our field is in transition. Until fairly recently, research in undergraduate mathematics education was almost exclusively concerned with observing and documenting the difficulties people have in learning college mathematics. We may call this the *observation/assessment* phase. Its implicit goals are to define and refine the problems. Very few papers attempted with any success to develop data–related theoretical perspectives to help understand the difficulties, that is, to interpret the data. The *interpretation* phase had hardly begun. Essentially absent were any signs of a *response* phase, that is, papers that reported success in making a difference through instructional strategies that were based on interpretively–linked data and theory.

We hope to illustrate in this book a new dynamism in which our field is moving from an entirely observation/assessment set of activities into more interpretive studies. Moreover, we are beginning to explore ways in which we can develop responses in the form of theoretically driven and effective instructional strategies. It is in the sense of these changes that we say that the field of research in undergraduate mathematics education is in transition from observation/assessment to interpretation to response.

Our book illustrates a movement through observation/assessment, interpretation, and response phases in two ways. First, the paper of Becker and Pence takes a long look at research reports which have appeared during the 16-year period from 1975 to 1991, and we can see indications of the transition in these papers. Second, we can situate the other eight papers in our collection within one or another of these three categories. It is perhaps useful to look at the papers in a little more detail with reference to this theme.

The majority of papers which Becker and Pence describe are of a mainly observational nature, but there are several which attempt to develop theoretical perspectives to explain the observations. Although there are also quite a few papers which describe innovative teaching methods to deal with the difficulties reported in the observational papers, there are very few examples of teaching methods based on theoretical perspectives.

For example, Becker and Pence list all of the national and international assessments and national reports that document the ways in which students have serious difficulties in learning mathematics. They also list papers which describe misconceptions and general difficulties with concept formation, as well as gender differences in mathematics achievement and participation, and the alarming situation with respect to most minority groups. We can see serious attempts at developing theoretical explanations of these difficulties in papers that investigate such topics as different learning styles, relational versus instrumental understanding, multiple -linked representations, and reflective abstraction. To be sure, Becker and Pence report on a very large number of curriculum development projects and innovative teaching experiments using technology, cooperative learning, self-paced instruction and other methods. It seems clear, however, that there is a serious lacking in the connection between these instructional strategies, and the developing theoretical perspectives. In our view, the task of closing this gap should be a major item on the agenda for research in undergraduate mathematics education.

The remaining eight papers in this collection suggest that this item is, indeed, finding its way onto our agenda. Only one paper, that of Selden, Selden and Mason, in which they look at the ability of "good" calculus and students to solve non–standard calculus problems, is strictly an assessment of the status quo.

The next four papers describe, in addition to many observations, efforts to develop theoretical explanations of various student difficulties. Ferrini–Mundy and Graham rely on a constructivist perspective to guide their methodology and analysis of data. They assume that, in using their previous experience to make sense out of mathematical situations, students make mental constructions that are rational and subject to explanation. They consider resulting errors and misconceptions to be anticipated phenomena that are natural in the learning process. Hart develops a perspective that is analogous to the van Hiele levels model that has received so much attention in geometry. In looking at content domain experts, he considers various levels of "expertise" and compares student responses in terms of these levels. Norman and Prichard adopt a Krutetskiian perspective of problem–solving and attempt to relate it to Cocking and Chipman's model of cognitive obstacles. They use this synthesis as the basis for their analysis of student difficulties. Finally, Rosamond looks at the types of emotional reactions that occur during mathematical problem–solving and tries to identify ways in which emotion inhibits or enhances learning. In this early foray into unexplored territory, she tries to obtain insight into problem–solving abilities by comparing the emotional reactions of experts and novices.

Turning to the third phase of our transitional theme, we can identify three papers which are concerned with instructional strategies based on theoretical perspectives. The papers of both Baxter and Cuoco describe the results of learning experiences using instructional treatments based on a Piagetian constructivism. In this perspective, learning mathematics occurs through construction of actions, interiorizing these actions to obtain mental processes, and encapsulating processes to construct mathematical objects. A major aspect of their point of view is that, after mathematical topics are analyzed in terms of such constructions, it is possible to design computer activities that will directly stimulate students to make them, first on the computer and then in their minds. These two papers offer a variety of computer applications that differ from each other in important ways.

A completely different point of view is taken by Bonsangue, who reverses the relationship between theory and practice. Rather than trying to derive an instructional strategy from a theoretical perspective, he takes cooperative learning, presently being used by many people in mathematics education, and tries to develop a theoretical foundation for it. He does this by conducting a traditional control/experimental–group–with–statistical–analysis study of the effect of cooperative learning.

OTHER THEMES

Another important theme illustrated by the papers in this collection is that researchers in undergraduate mathematics education are operating on a broad front. Aside from the survey of Becker and Pence, the eight remaining papers have to do with six mathematics topics. Baxter's paper is about the concept of sets, and Cuoco's paper is about functions. It is not surprising, given today's calculus reform movement, that three papers, those of Ferrini–Mundy and Graham, Norman and Prichard, and Selden, Selden, and Mason are about Calculus. Bonsangue studies Statistics, Hart investigates Abstract Algebra, and Rosamond reports on the affective dimension of general problem–solving skills in mathematics.

Finally, the papers display the variety of research paradigms accepted and used in the field. Ferrini–Mundy and Graham are content to look at a single student and closely analyze the development of that student's understanding. Baxter and Cuoco, in a similar spirit, each look at several individual students. Both Norman and Prichard and Selden, Selden, and Mason study the knowledge and understanding of traditional Calculus students. All of these authors are trying to understand what they observe, leaving comparisons as an "exercise for the reader." Bonsangue, on the other hand, makes a careful and traditional statistical comparison of two classes, each as identical as possible except in the use of cooperative learning. Hart tries to compare the ability to make proofs on the part of students from several levels of expertise, and Rosamond compares the emotional content of problem–solving experiences of novices and experts.

CONCLUSION

We hope that the papers in this collection will help demonstrate that Research in Undergraduate Mathematics Education is a young, healthy, and dynamic field. We hope that some readers will find things here that help with classroom teaching, or at least suggest a potential for such help. We hope that some readers will be sufficiently excited by what is offered here and elsewhere, including the many references, to think seriously about getting involved in doing research in this area. Most of all, we hope that readers will accept the idea that this field is becoming a valid area of scholarly work for those of us who are trained in mathematics and are experienced in teaching, as well as those of us who have formal backgrounds in educational research. For the former, it will require an effort to become familiar with the literature and to learn how this research is done. It is very different both from doing mathematics and preparing for teaching a class. For those with experience in education research, it requires only a movement into areas of advanced mathematical thinking.

As we indicated at the beginning, post–secondary mathematics education, like most areas of learning mathematics, is in deep trouble. We hope to demonstrate in this book that, research, in addition to being important as a scholarly endeavor, can be part of the solution to our educational problems.

Acknowledgment: We wish to thank Audrey Chekares for chasing down and taming unruly revisions and to thank Richard Faulkenberry for creating and managing the technology that enabled her to succeed.

Jim Kaput
Ed Dubinsky
May, 1993

Contents

PART I

LITERATURE

BECKER AND PENCE — COMMENTS

When a new field of scholarly endeavor arises, or when a very small "out–of–the–way" area begins to attract attention, grow, and even aspire to a certain prominence, and when that field considers the dissemination of research reports to be one of its prime concerns, then a certain number of "growing pains" are bound to occur. First, there is not enough research and consequently, not very many reports. Fledging journals have trouble filling their pages and their survival is threatened. After some time, if the field survives, research activity grows and all of a sudden there are too many papers. Or, as may be the case with research in undergraduate mathematics education, a growing number of people are engaged in research, or in activities close to research, but their work is hindered by the scarcity of outlets for publication. At that point, if the natural forces causing the growth of the field are strong, the number of research reports increase, and they get published in journals for which they are an exception, or in collections like the present volume, or they are just circulated informally among professional acquaintances.

There is reason to believe that this is what is happening with the field of research in undergraduate mathematics education. More work is being done than gets formalized as "research", but there is still a lot being published — much more than the few journals in mathematics education can handle. This is especially so since those journals devote most of their pages to research in K–12 mathematics education. The rest of the work appears as exceptional papers in various journals, or in the annual Proceedings of International Meeting of Psychology of Mathematics Education (PME), or in other conference proceedings.

Aside from the problem of stimulating even more research and the resulting publications, the present state of affairs is that a fairly large and growing number of papers (Becker and Pence found approximately 650 appearing in the period 1975–1991, and about 150 of these in the last two years) are published in a variety of places and there is very little in the way of quality standards, post–publication review, and some sort of order to help the individual who is trying to get a grasp of what is going on. There is very little to guide the novice, or even the expert, who would like to include a literature review in a paper.

This situation is a natural consequence of something that we can welcome — the rapid growth of our field. We can certainly live with it for a while. Eventually, however, some organization will have to be introduced into the chaos. Ultimately, for example, we will need a review journal. For now, however, it suffices that there are a number of individual and specialized efforts. The Journal for Research in Mathematics Education publishes an annual index which, although it covers all levels, has enough structure to allow the reader to concentrate on post–secondary work. There are a number of individual efforts such as the paper of Leinhardt et al, which reviews the literature on the concept of function.

The first paper in the present collection is such an effort extended to the entire field of research in undergraduate mathematics education. Becker and Pence took the 650 published papers that they found and selected 165 of them to review. Their report, however, does not consist of a collection of reviews of individual papers. Rather, what they have done is to establish a number of categories and then, for each category, they summarize what can be found in the totality of the papers they reviewed. In this way, we get a synthesis of what appears in the field and what the authors think about it all.

This is as it should be. A good review should not confine itself to presenting a digest of the work being reviewed or the author's abstract, but rather should be a vehicle for the reviewer(s) to express opinions, make selective judgments, and propose structures for the field. This is very much what Becker and Pence have done in their paper. They consider five categories: Student learning, Teaching methodology, Equity issues, Academic preparation for college, and Teacher education. Several of the categories have sub-categories. Thus, for example, under Student learning, we see learning styles, problem–solving, concept formation and specific mathematical topics. Equity issues includes subcategories concerned with gender and with minorities.

As we said above, the selection reflects the authors' judgments, and so the reader should not be surprised if the emphasis given to this or that topic is greater or smaller than expected. Moreover, although the paper is very current regarding work published through 1991, there is the usual one-to-two year lag between a report and its publication. As we noted above, about 150 papers have been published in the last two years, so at least that many must be in press. To this must be added the time of getting the present collection "out the door." As a result, any work of this kind must be somewhat out of date from its beginning.

There are some manifestations of the fact that a review paper cannot reflect the very latest work. There are in press, in various publications, a number of papers describing various theoretical perspectives that guide the editors' research. Becker and Pence will not have seen many of these in print, and as a consequence this topic is almost completely missing from their review. The same can be said about mathematical subjects. Here we have Calculus and High School algebra. Papers in press or in preparation will expand our areas of interest to include Differential Equations, Linear Algebra, and Abstract Algebra.

It is also possible to make some predictions. For example, it seems likely that the near future will see research papers, at the college level, on cooperative learning and methods of evaluation of learning.

If the field of research in undergraduate mathematics education is not actually a supernova, it is certainly producing a rapidly expanding body of work. We cannot expect a review paper to bring us right up to date. What we can expect, and what Becker and Pence do achieve for us, is to be made relatively current by reading such a paper — up to two or three years before its appearance. The serious investigator then has to fill the gap with a comprehensive survey of all relevant publications, but for a much shorter time period than would have been necessary were it not for papers such as this.

Finally, we reiterate that Becker and Pence, by their categorization, propose a structure for our field. We can accept it, revise it, or substitute something entirely different. But what they have given us is a start.

THE TEACHING AND LEARNING OF COLLEGE MATHEMATICS: CURRENT STATUS AND FUTURE DIRECTIONS

Joanne Rossi Becker and Barbara J. Pence
San Jose State University

This paper is a shorter version of a paper and bibliography undertaken on request from the California State University Institute for Teaching and Learning to review the state of knowledge about college mathematics. Because we found a compilation of sources covering the years 1900 to 1974 (Suydam, 1975), we decided to concentrate on materials published since 1975. Because of the limited time in which to complete the project, we cannot claim to have done an exhaustive search. We used primarily the July issues 1975–1991 of the **Journal for Research in Mathematics Education**, which contain a list of journal articles and dissertations published in the previous year. These compilations [see, for example, Suydam, 1989] are helpful because they not only provide a brief description of the research, but also provide the grade level(s) involved. Thus we were able to identify all sources pertinent to the college level. The research summaries search a broad selection of journals in education, mathematics education, and mathematics, but of course may not be exhaustive. Journals which are indexed in the July issues of the **Journal for Research in Mathematics Education** were searched issue by issue for the 1991 calendar year.

For this review we concentrated on published manuscripts. We searched ERIC documents to look for other reviews of college level sources. Besides the Suydam (1975) bibliography, we found one other ERIC document which reviewed all research in mathematics education in 1987 (Dessart, 1989); this included a section on college level research. In addition to these sources, we also searched page by page through the proceedings of the International Group for the Psychology of Mathematics Education, 1985 through 1991. Reports of professional groups and other books of interest have been included as well.

As a result of these searches, approximately 650 sources were reviewed for the full paper and bibliography. Because of space limitations, we have selected for review in this paper those references which we judged added the most to our knowledge about teaching and learning mathematics at the college level.

PAPER ORGANIZATION

The paper is organized into three major sections: Need for Research on Teaching and Learning College Mathematics; Current Status of Knowledge on Teaching and Learning College Mathematics; and Future Directions. The first section will establish the current active state of calls for reform in collegiate mathematics education. The second section will summarize the research we identified on teaching and learning college mathematics. And the third section will present our view of needed research in the future.

NEED FOR RESEARCH ON TEACHING AND LEARNING COLLEGE MATHEMATICS

BACKGROUND

The recognition of the need for reform in mathematics curriculum and instruction is broad and deep, ranging from professional organizations (Douglas, 1986; Steen, 1989; National Research Council, 1989, 1991; National Council of Teachers of Mathematics (NCTM), 1989) to government agencies (The National Science Board Commission on Precollege Education in Mathematics, Science and Technology, 1983; The Task Force on Women, Minorities, and the Handicapped in Science and Technology, 1988, 1989) and business and industry leaders (Johnston & Packer, 1987). The call for reform is remarkably consistent across the grade levels K–college, and has stemmed from several sources, including the underachievement in mathematics by US students (Dossey, Mullis, Lindquist & Chambers, 1988; McKnight, Crosswhite, Dossey, Kifer, Swafford, Travers & Cooney, 1987); the decline in numbers of young people studying mathematics, the sciences, and engineering (The Task Force on Women, Minorities, and the Handicapped in Science and Technology, 1988, 1989); the changing demographics of the school population and the future workforce (Johnston & Packer, 1987); changing perspectives on the nature and learning of mathematics (National Research Council, 1986, 1989); and the impact of technology on education and society. **Everybody Counts** (National Research Council, 1989) eloquently summarizes these challenges to the future of mathematics education, and we refer the reader to that document for further detail.

COLLEGE LEVEL CURRICULUM REFORM

According to **Everybody Counts**, "Reform of undergraduate mathematics is the key to revitalizing mathematics education" (National Research Council, 1989, p. 39). If reform of undergraduate mathematics is the key, surely reform of calculus is a critical first step because of its unique position as the bridge from high school to college mathematics. Although a variety of courses are recommended for the first two years of mathematical sciences in college (Steen, 1989), calculus is still considered a major part of those years (Douglas, 1986; Steen, 1987), and much of the curriculum revision to date has focused on the calculus. Conferences and reports have revitalized interest in calculus reform, and the National Science Foundation (NSF) has recognized its importance by initiating the Undergraduate Curriculum Development Program in Calculus. In 1988 NSF awarded five multi–year grants, one of which provided startup money for **UME Trends** (Dubinsky, 1989), 19 one–year grants, and funds for a series of conferences. **UME Trends** has had several articles outlining the other four multi–year calculus projects (see for example, Jackson, 1989a, 1989b), and is a good resource for continuing information on calculus. In 1989, NSF awarded 17 additional grants. Funded projects range from: those which aim to enrich the traditional course without changing the syllabus, for example by using a new technique such as student research projects; to projects which aim to integrate use of calculators and computers into the curriculum, with an accompanying emphasis on problem–solving and de–emphasis on manipulative skills; to projects which integrate calculus with physics; to projects which attempt to restructure the entire course so that concepts arise from real life problems in the social, life, and physical sciences. Certainly much work is also being done at sites without funding; the majority of efforts seem to focus on making use of technology to refocus the course and change the instruction (see, for example, Bauldry, 1990; Beckmann, 1990; Berenson & Stiff, 1990/91; Fraleigh & Pakula, 1986; Heid, 1988; Mathews, in press; Palmiter, 1990; Zorn, 1990). However, to date there has been very little published research or evaluation concerning these new initiatives (see Calculus section below). Careful documentation and research are essential if we are to judge what specific strategies and approaches further the reform goals specified in **Toward a Lean and Lively Calculus** (Douglas, 1986) and **Calculus for a New Century** (Steen, 1987). Curriculum decisions need to be made on more than philosophical bases ungrounded in research.

CURRENT STATUS OF KNOWLEDGE ON TEACHING
AND LEARNING COLLEGE MATHEMATICS

This section will review the selected research directly pertaining to teaching or learning mathematics at the college level. Although we identified a large number of published reports of such research, overall we must say that the research seems to be sadly lacking in usable knowledge. Much of the research that has been done at the college level suffers from theoretical and methodological deficiencies which make it hard to generalize the results to any population beyond that immediately sampled. In each of the following sections, review of the current status of research–supported knowledge will be supplemented by the identification of promising directions for future research.

STUDENT LEARNING

This section will deal primarily with cognitive aspects of learning; affective issues will be discussed in the section on Equity Issues and Access to Mathematics.

Learning styles. Recent research on learning styles identifies at least two distinct styles of learning: splitters, who tend to analyze information logically and break it down into smaller parts; and lumpers, who tend to watch for patterns and relationships (Claxton & Murrell, 1987). There may be gender differences in preferred style. In **Women's Ways of Knowing,** (Belenky, Clinchy, Goldberger & Tarule, 1986), the authors reexamine Perry's scheme of intellectual development in college students (Perry, 1970) with a sample of women students and community members. The authors found that, for many women, the subjective stage of knowing which emphasizes the intuitive, personal approach, relationships, and discovery, is a critical one in their intellectual development. Attributes of this stage are incorporated in connected knowing, in which authority comes from shared experiences from which a conclusion can be made. In the constructed knowing perspective, all knowledge is constructed by the knower and answers are dependent on the context in which the question is asked. The knower integrates rational and intuitive thought and can appreciate the complexity of knowledge from various perspectives. The predilection of women for connected knowing may be in direct conflict with the way in which much of mathematics is taught.

Buerk (1985) provides some further insight into this topic. She found that a dualistic view of mathematics as consisting of only rules, formulas, and skills, with all questions having answers known to an authority, was related to mathematics avoidance. She also found that a change in the conception of mathematics could help students develop more complex thinking and lead to more positive attitudes. Buerk (1985) includes suggestions for specific strategies which enhance mathematics learning, including encouraging collaboration, intuition, and questions, avoiding absolutist language, providing time to experience and clarify problems, and acknowledging alternative methods of solution. She also advocates including an historical perspective to give the view of mathematics as human–made and to illustrate that many people struggled for years with difficult concepts before they were accepted. Buerk (1985) uses writing to allow students to reflect on their ideas and feelings concerning mathematics; this seems to provide an opportunity for negative feelings to be vented and thus diminished.

There are some indications from research with precollege minority students (Valverde, 1984) that there is a specific cognitive style which fits many Hispanics, namely field dependence. Field-dependent learners may learn mathematics more readily when the teacher provides a high level of guidance and structure to the lesson. These students seem to find materials with social content more attractive, and find working together with minimal competition more conducive to learning.

Dunn, Sklar, Beaudry, & Bruno (1990) provide more evidence that learning style interacts with instructional styles. In a study of over 1,100 students of diverse ethnicity, they found that simultaneous processors achieved significantly better when taught globally while successive processors achieved better when taught analytically. Perhaps more connected teaching as Buerk (1985) describes would help a more diverse population of students of both genders to succeed in mathematics.

The role of visualization in learning mathematics is an open and important question (Harel, 1989a; Vinner, 1989) and relates to learning-style differences. Lean and Clements (1981) found a tendency for students who preferred to process information by verbal–logical means to outperform those who preferred a visual presentation on both mathematical and spatial tasks. However, Eisenberg and Dreyfus (1986) compared research mathematicians, high school mathematics teachers, and university students on performance on eight problems which were stated in symbolic form but provoked both visual and analytical answers. They concluded that cognitive orientations vary, visualization is not a prerequisite for the study of higher level mathematics, and both analytical and visual aspects need to be presented in instruction. An interesting study by Harel (1989b) of a beginning linear algebra course varied the embodiments used and the mode of representation of the concept of vector space. Familiar geometric embodiments as compared to unfamiliar algebraic ones were found to be superior. Familiar geometric representations seemed to provide a significant contribution to the formation of the concept of vector space. However, Harel did find students who preferred a geometric mode even if they had not been exposed to it, and students who had been exposed to the geometric content preferring an algebraic mode of solving problems.

In summary, recent results in intellectual development seem to point to different styles with which instructors should be familiar so that they can design learning experiences which match, or mismatch, students' styles, depending on the instructor's intentions. Clearly it would be preferable to find instructional styles which are inclusionary rather than exclusionary, so that students with different styles can still learn effectively. Learning styles is a major area in need of further research. In particular, we need to identify how much difference it makes if teaching methods are incongruent with a student's preferred style. More research on the learning styles of women and minorities and their effect on learning college mathematics seems critical as we target these groups for careers in science and technology. Hopefully such research would seek ways to teach mathematics so that all students could.

Problem Solving. Mathematical problem–solving has experienced a resurgence of interest in the last decade or so. As evidence, note three recent volumes relating to problem–solving: Charles and Silver (1989); Schoenfeld (1983a); and Silver (1985a). The Schoenfeld report arose from a survey conducted by the MAA on problem–solving courses nationwide. It includes Schoenfeld's suggestions for teaching mathematical problem–solving, and a valuable bibliography.

Research on problem–solving has been influenced by work in cognitive psychology, artificial intelligence, and affect. Qualitative methodologies are most often used in problem–solving research; researchers are particularly concerned with the process of problem–solving which can only readily be ascertained through methodologies such as interviews, thinking aloud, or teaching experiments.

Without instruction, college students are rather poor problem solvers. Eisenberg and Dreyfus (1985) had students work on six problems; their solutions were classified by paths and patterns used and whether an elegant solution was considered. It was found that students' backgrounds, courses taken, and grades could be ignored, because all students rushed towards an answer, used known procedures uncritically, seldom questioned whether alternative solutions were available, and did not generalize unless asked to do so. Each problem was approached as a separate entity, with little perception of similarity in process or problem-structure.

There is growing evidence that students have difficulty sorting problems by their mathematical structure (Gliner, 1989, 1991; Reed, 1989). More successful problem-solvers are able to recognize the structure of problems; the less able tend to sort problems by context, question form, or common units. But merely directing students' attention to similarities between problems does not seem sufficient to improve problem–solving performance. Bassok (1990), in a study of transfer, found a complex relationship between content and structure of problems; many features of content may be screened out, but content features needed to interpret variables may affect transfer from one domain, such as physics, to another, such as algebra.

Graduate students in mathematics did not demonstrate strong problem–solving skills in Trelinski's work (1983). Participants were given a situation outside of mathematics and asked to create a mathematical model for its solution. Of 223 subjects, only nine were fully successful. Subjects had a tendency to make global descriptions that did not account for problem details, and disregarded the importance and impact of assumptions they made. On the whole they showed an inability to identify mathematical concepts in situations outside of mathematics; their problem–solving skills in mathematics did not transfer to other disciplines.

Considerable evidence is available demonstrating that instruction in use of heuristics and a general process for attacking problems can be helpful to students' development as problem-solvers. For example, Schoenfeld (1979) found that students taught heuristics outperformed other students on similar problems. Cope and Murphy (1981) also found that use of strategies could be taught and was necessary for the solution of problems involving organization of well–understood parts. Schoenfeld (1982) identified two conditions for success in problem–solving: mastery of basic problem–solving techniques, and a managerial strategy which helps one budget problem–solving resources efficiently. A metacognitive perspective which allows the student to reflect on her/his problem–solving work is also useful.

Affective factors have also been identified as important components of problem–solving ability (McLeod, 1985; Schoenfeld, 1983b). Successful problem-solvers are not only persistent and confident, but also have an open view of the nature of mathematics. Change in affective variables is difficult and time-consuming, but possible (Thompson, 1989).

Researchers in problem–solving hold a vision of mathematics classrooms as situated, collaborative environments involving social construction of knowledge and socially–distributed problem–solving (Silver, 1989). What is needed is more guidance on how to create and nurture such an environment. Research is needed to help develop effective strategies to improve the teaching and assessment of problem–solving in classroom settings.

Concept Formation and Misconceptions. Students bring to mathematics classrooms a set of intuitions and partial understandings. This knowledge is restricted. Errors and misconceptions develop as students extend and process their knowledge in new contexts. The most valuable studies grouped in this category used qualitative methodologies such as clinical interviews, which provide a depth not available in paper–and–pencil tasks.

As the reader is well aware, even college students have misconceptions about basic mathematical concepts. For example, Grossman (1983) found that only 30 percent of entering students could correctly identify the smallest decimal number in a finite list, although over 50 percent could correctly operate with decimals. Even prospective teachers lack depth of understanding of basic concepts such as division; Ball (1990) found that prospective teachers were able to solve problems mechanically but could not construct meaningful explanations.

Function is a critical concept which causes much difficulty and is presently the focus of serious study (Harel & Dubinsky, 1992). Students seem to have a restricted concept of function, with little connection developed between the symbolic representation and the graph (Bakar & Tall, 1991; Even, 1988; Goldenberg, 1987; Harel & Dubinsky, 1991; Smith, Arcavi & Schoenfeld, 1989). Computers provide a unique tool to investigate how multiple-linked representations of functions might be developed and how such linkages contribute to concept and schema development (Goldenberg, 1987; Heid, 1988, 1990; Kaput, 1989). Students seem to understand function mainly as a formula; Ferrini–Mundy & Graham (1991), using clinical interviews, found that students could not provide a general definition of function but could give examples by writing formulas. They conclude that students do not see functions as objects of study in mathematics, but rather as equations with which one is expected to do something, such as substitute a value.

The formal definition and the concept of a function may be inconsistent in the mind of the student. Vinner and Dreyfus (1989) found that many inconsistencies between the concept and definition existed. For example, 56 percent of all respondents who gave the Dirichlet–Bourbaki definition for function did not use the definition when justifying answers given relative to the examples.

Misconceptions abound at all levels. Artigue (1986), when attempting to coordinate courses in mathematics and physics, found that "differential" had different meanings to both students and teachers and played a different role in the teaching. Two conceptions of differential coexisted in students and were linked to the different contexts, mathematics or physics, but apparently not linked to each other. Tall and

Schwarzenberger (1978) and Tall and Vinner (1981) investigated difficulties in learning the limit concept. They found that student difficulty may be caused in part by the informal translations provided of the formal definition, for example, by saying that a sequence S_n has a limit S if we can make S_n as close to S as we want by making n sufficiently large. This may lead students to think that S_n can be close but not equal to S; thus Tall and Schwarzenberger (1978) found that many undergraduates thought that 0.999... is less than 1.

Williams (1991) also investigated students' conceptions of the limit concept and factors affecting change in understanding. Although problems he presented were chosen to encourage change to a more formal conception of limit, the dynamic aspect of students' models was very hard to change. Students were not motivated to adopt a formal view of limit.

More evidence of the deficiencies of surface structures is found in Amit and Movshovits–Hadar (1991), Amit and Vinner (1990), Even (1990), Hansen, McCann and Myers (1985), Mayer (1977), and Hardiman (1984). Each study supports the importance of relational learning and the quality of the schema to understanding and transfer. Skemp (1986) provides a good background on these concepts. More recent work by Harel and Kaput (1990) provides a discussion of these ideas as they relate to mathematical concepts from the undergraduate curriculum.

Misconceptions about proofs and inabilities to construct proofs are widespread at the high-school level, but also have been found to continue into college and beyond in studies such as Hitt (1989), Martin and Harel (1989), Moore (1991), and Lewis and Anderson (1985). Proof is the topic of other articles found in this volume as well. Closely related are studies of students' logic and reasoning. Studies to classify college students on Piaget's developmental stages identify a substantial number of college students as concrete operational thinkers (Nummedal & Collea, 1981; Thomas & Grouws, 1984; Thornton & Fuller, 1981). Piaget's concept of reflective abstraction also motivated Dubinsky (1988) to propose a theory of knowledge and its acquisition and to investigate both induction and quantification in light of this theory.

Much of the work cited above focuses on misunderstandings and partially formed ideas. Many misconceptions abound, and it seems that we need much more development of students' intuitive notions before we introduce abstract concepts. More research is needed to develop models of understanding and to perfect our methods of assessing understanding.

Algebra. Although much of the research on student learning of algebra has been done with secondary school students, a substantial part of mathematics teaching at two–year and four–year colleges includes high school algebra. Therefore we felt that some discussion of this topic was necessary. To many students, algebra presents the problem of learning to manipulate symbols according to certain transformation rules, that is, syntactically, without reference to the meaning of the expressions or transformations, that is, the semantics. While knowledge of the cognitive aspects of learning algebra is compiled in Wagner and Kieran's monograph (1989), the pedagogical question of how we should teach algebra so as to facilitate its learning remains a challenge for further research. In particular, the role of technology, with its ability to provide new representations and new possibilities for multiple linkages among representations (a web), is a new, potentially fruitful avenue of research (Kaput, 1989).

Calculus. In the "Need" section we discussed some of the curriculum initiatives concerning calculus. Here we review the small amount of published research concerning the calculus.

Because the use of computers in the calculus is quite new, there is little published research evaluating the effects of that innovation. The research available does point to an improvement in conceptual understanding or higher level thinking. For example, Heid (1988) resequenced an applied calculus course by using the first 12 weeks of the semester to study concepts with graphical and symbol–manipulation computer programs to perform routine manipulations. Only the last three weeks were spent on skill development. Through a variety of measures, Heid found that students in the experimental classes showed better understanding and almost as high skill performance as a class which had practiced skills for the entire 15 weeks. Beckmann (1990) used four treatments to investigate computer graphic use in first semester calculus. The graphics groups performed better on non–routine symbolic questions, but the traditional skill–oriented group performed better on routine symbolic questions. Retention was much higher in all conceptually-oriented sections (computer and non–computer) than the skills section. Student attitudes toward use of graphics was very positive. Tall (1985), investigating the use of computer graphics in teaching the derivative concept, found that the geometric notion of the derivative was enhanced. Bauldry (1990) reported on a pilot use of a one–hour computer lab per week in conjunction with a calculus class, in which students primarily used a graphing package. Bauldry found that students' graphical perception was greatly enhanced, and geometric intuition developed in the courses was superior to that of non–computer students. And Palmiter (1991) found that students using a computer algebra system in calculus scored higher on a test of conceptual knowledge than those taught in a traditional manner. These students also did better on a computational exam using the computer algebra system than did their peers using paper and pencil only.

With all the curriculum development work on calculus currently under way, there is a critical need for more careful study of specific innovations such as these few cited. But certainly we have sufficient indications of the positive effects that the use of computers can have on the teaching of calculus to encourage more experimentation.

Technology. The impact of technology on the teaching of college level mathematics is receiving considerable attention, as noted above in discussing revisions in the calculus. An excellent overview of the potential impact can be found in Ralston (1990). Research findings are just beginning to be published; we can expect this to be a major area of research for some time to come as people grapple with the best ways to use the powerful technology now available. Contained within our publication are several such developing studies.

Past research involving computer–assisted instruction has been quite mixed in results. Newer findings, using more sophisticated software, show more promise. For example, both Sterling and Gray (1991) and Sasser (1990/91) found that the use of computers improved mathematics achievement significantly.

An interesting line of research is opened by Browning (1990), who discusses the development of an instrument for evaluating students' levels of graphical understanding. Such instruments can be very helpful as we undertake research to ascertain the impact of instructional technology. Browning designed a "Graphing Levels Test" for precalculus students. This instrument was formed in levels in a developmental sequence similar to the van Hiele levels of learning geometry: Level 1: recognition of graphs, initial vocabulary; Level 2: translation of verbal information into a simple sketch of a graph, and simple interpolation from a graph; Level 3: usage of properties of graphs of functions to determine functions from non–functions, recognition of the connection between a graph and its algebraic representation, and usage of properties of functions to construct graphs; Level 4: usage of given information to construct a graph, usage of information from a graph to deduce more information, and recognition of what is required to find from given information. Comparing a computer/calculator precalculus with a traditional class, Browning found that 69 percent of the control group was at Level 2 or below at the end of the class, but 73 percent of the experimental group was at Level 3 or 4. Browning concluded that the technology provided students with more examples and aided them in making mathematical connections.

Additional evidence to support use of computers to develop understanding is provided by Ayers, Davis, Dubinsky, and Lewin (1988). They used a computer to help students learn the concepts of function and composition of functions. Those students who had computer experiences scored higher on a test of understanding of these concepts than a control group taught in the traditional way. Ayers et al conclude that the computer can be used to provide instantiations of abstract concepts, and thus help induce reflective abstraction, a Piagetian cognitive process through which physical or mental action is reorganized on a higher plane of thought.

Work by Kiser (1990) alerts us to the possibility that students with strong spatial skills may benefit more from work in a computer–enhanced environment. More study is needed of this issue.

Finally, a new use of technology which has had little study is in testing. Fetta and Harvey (1990 a, b) discuss the ways in which technology is changing tests and testing practices; technologically–based systems have administrative advantages, but we need a better understanding of the psychological effects of computerized testing on grading. Hsu and Sheemis (1989), using a computerized placement testing system for college mathematics, found that students preferred that system over pencil/paper testing.

Integration of technology provides exciting opportunities, but also poses new questions about student learning in this technological context which are just beginning to be explored.

SUMMARY. Overall, the research presented focusing on cognitive aspects of student learning includes some new perspectives on learning style, a substantial amount of information about student errors and misconceptions, a reasonable set of conclusions concerning mathematical problem–solving, and exciting possibilities for technology to enhance conceptual understanding. Much of this research has implications for pedagogy. In the next section we examine the sources we identified concerning teaching methodology.

METHODOLOGY

The division between student learning and methodology was not always clear; references were put in this section if they were judged to be primarily discussing the efficacy of a particular method of teaching.

Effective Teaching. A number of studies have investigated teacher clarity (Hativa, 1983, 1985; Land & Smith, 1979; Smith & Edmonds, 1978; Smith & Land, 1980). Overall, these studies have identified characteristics of teacher clarity as measured by observers' or students' ratings, and found that teacher clarity consistently and significantly correlates with student achievement. Hativa (1985) classified teaching characteristics pertaining to teacher clarity into three areas: those which make a lesson easy to follow, such as structuring the lesson through outlines; those which make a lesson easy to understand, such as helping students connect new material to old; and those which make a lesson easy to remember, such as identification of important points through emphasis and summaries. Teachers who are not clear use vague terms, start a sentence without completion, hesitate, skip intermediate steps, state facts students cannot yet understand, make errors in computation or proofs, and make wrong or inaccurate statements. These attributes of clear/vague teaching relate specifically to a lecture presentation style.

An interesting direction to take in identifying effective teaching is illustrated by Rogers (1988), who studied the State University of New York at Potsdam, well known for the large number of mathematics degrees awarded. Through qualitative methods, Rogers identified characteristics of the department which were related to its success. These included a learning environment imbued with a culture of success, emphasis on all students' learning to think mathematically through student–centered teaching methodologies instead of lectures, and a flexible grading and assessment scheme. In summary, they use a student–sensitive, constructivist approach to instruction.

Tests and Assignments. Although many instructors probably experiment with various approaches to testing and homework, we found very little published research on these topics. If homework is collected, the sooner feedback is given to students the better (Austin, 1980). And Hirsch, Kapoor and Laing (1983) compared distributive homework [half on new material, half review] with homework related to daily topic only in a first semester calculus class. Students who had weaker mathematics backgrounds profited more from the distributive homework, while those with stronger backgrounds achieved better when the homework was related to daily topics only. This finding suggests that perhaps differentiated assignments might be helpful.

One study of the use of retesting (Deatsman, 1979) found that retesting harmed slower students who used it as a rehearsal and did not adequately prepare for it. Anderson (1989) investigated gender differences on objective tests with a sample of mathematics majors; he found that women averaged about 80% of men's scores on multiple choice and true–false questions even though they performed equally well on standard open–ended questions. The women seemed less likely to guess. Thus it would seem that objective tests should be used with caution.

Cooperative Learning. Many reform reports call for more collaborative teaching methods in mathematics at all levels. Small group learning was one method used at Potsdam (Rogers, 1988). Brechting and Hirsch (1977), Chang (1977), Dees (1991), and Shaughnessy (1977) all found small group methods more effective in college classes. Davidson (1990a) provides a lot of useful information about cooperative learning; in particular, the chapter by Davidson (1990b) specifically relates to using the method with college-level students, and includes guidelines, sample mathematical situations for group discussion, and a review of research findings concerning that method. He summarizes the research by stating that students taught by the small–group discovery method have performed at least as well as those taught by more traditional lecture methods. Advantages of cooperative groups include: active student involvement; opportunity to communicate mathematically; a relaxed, informal classroom atmosphere; freedom to ask questions; a closer student–teacher relationship; high level of student interest; more positive student attitudes; and opportunity for students to pursue challenging mathematical situations.

An additional reason to consider cooperative group learning is its potential to enhance the learning of women and minority students (Cole & Griffin, 1987; Fennema & Meyer, 1989) However, we caution that instructors should watch for differential treatment and participation within groups, and not assume that all group participants are experiencing the higher level thinking, active engagement, and communication which groups may foster.

SUMMARY. The research on methods of teaching mathematics at the college level leads to some generalizations but also to many more questions. Specific teaching behaviors related to clarity lead to better student achievement and higher teacher ratings. Cooperative learning is a promising alternative to the traditional lecture method which merits wider use and investigation. Certainly, as we move toward implementation of the recommendations for change in mathematics instruction, there should be many more research questions relating to methodology.

EQUITY ISSUES

Because many of the references concerning affective factors influencing learning relate to issues of equity, we have put all references on affect in this section.

Affective Variables. McLeod (1987) provides a good background for understanding the role of affective factors in mathematics learning, distinguishing among beliefs, attitudes, and emotions. McLeod and Adams (1989) provide a compilation of new research attempting to link affect and mathematical problem–solving, although much of the research does not deal with college samples. Most research on affect with college populations has involved anxiety as opposed to attitudes. In general, women tend to report lower attitudes toward mathematics and computers than men do (Aiken, 1976; Dambrot, Watkins–Malek, Silling, Marshall & Garver, 1985) and more anxiety about mathematics than men do(Adams & Holcomb, 1986; Cope, 1988). However, attitudes are clearly related to high-school mathematics or computer background (Dambrot et al., 1985; Stones, Beckmann & Stephens, 1983). Because women tend to have less pre–college background in both mathematics and computers, that may well account for some of the difference in attitudes and anxiety. Although the prevalence of mathematics anxiety may have been overblown in the popular press (Becker, 1986; Ohlson & Mein, 1977; Resnick, Viehe & Segal, 1982; Stones et al., 1983), it can be a significant stumbling block for both male and female students (Buxton, 1981). Anxiety may diminish women's achievements in mathematics, which in turn affects the pursuit of mathematics–related college majors.

To alleviate mathematics anxiety, a variety of approaches have been tried; these include study skills, cue–controlled relaxation techniques (Bander, Russell & Zamostny, 1982; Gentry & Underhill, 1987), mathematics instruction focusing on conceptual understanding (Buerk, 1985; Resek & Rupley, 1980), and psychological adaptive techniques (Gentry & Underhill, 1987). Study skills and replacement of maladaptive thoughts with more positive, rational ones produced reductions in mathematics anxiety. The math-anxious tend to be rule–oriented (Buerk, 1985; Resek & Rupley, 1980), and benefit more from an expository rather than a discovery approach (Clute, 1984). However, traditional expository teaching methods have helped produce current anxiety; special courses such as those described by Resek and Rupley (1980) and Buerk (1985) which focus on developing conceptual understanding in an emotionally supportive environment show the best promise for counteracting mathematics avoidance behavior.

Gender Issues. In the last 15 years there has been a considerable amount of research on gender–related differences in mathematics achievement and participation. Research about gender and mathematics might start by looking at the underrepresentation of women in scientific jobs in the US; although women represent 45 percent of the US workforce, they comprise only 11 percent of all employed scientists and engineers (NSF, 1988; Task Force on Women, Minorities, and the Handicapped in Science and Technology, 1988). Although more women are now pursuing undergraduate degrees in the mathematical sciences, they continue to comprise a relatively small percentage of those earning advanced degrees ("A Profile of 1987 Recipients of Doctorates," 1989). Much of the research in this area has focused on identifying reasons for this underrepresentation and has targeted pre–college populations. A review of that considerable research is beyond this paper; the interested reader might consult Fennema and Leder (1990). The research has investigated cognitive and psychosocial gender differences, as well as cultural and social factors. Two recent meta–analyses (Friedman, 1989; Linn & Hyde, 1989) should help dispel the notion that differences in cognitive or psychosocial variables can account for the underrepresentation of women; these studies found that gender differences on cognitive and psychosocial tasks are small and declining, and not, in any case, general. Thus we need to devote more attention to the effects of social institutions, particularly educational ones, on women's occupational choices if we want to increase the number of women in the scientific pipeline.

It is clear that women take fewer mathematics courses in college, and that even at that level mathematics may act as a critical filter keeping women from scientific careers (Boli, Allen & Payne, 1985; Deboer, 1984). College instructors can play an important role in encouraging women to pursue a mathematics–related career. Mura (1987) studied undergraduates and Becker (1984, 1989) graduate students in the mathematical sciences. All three studies identified confidence as a critical variable in the women's intention to pursue a Ph.D. The women expressed less confidence in their mathematical abilities, and thus needed more encouragement, particularly from college instructors, to pursue graduate studies. What affects the development of a student's confidence is not generally agreed upon, although we can hypothesize many causes. One theoretical model which merits investigation with college populations is attribution theory; with pre–college populations, it has been found that males tend to attribute their mathematical success to internal causes and their failure to external, unstable causes (Fennema & Meyer, 1989). Females, on the other hand, tend to attribute success to external or unstable causes and failures to internal ones. Thus females, more than males, show a pattern of attribution which is sometimes called learned helplessness. They may feel powerless to overcome failure because they attribute it to something they cannot change.

The contextual factors of education (Cole & Griffin, 1987) cannot be ignored in this discussion. One aspect of that context is the classroom climate. In a study of the classroom climate at the college level, Hall and Sandler (1982) classified it as a chilly one for women. Although not specific to mathematics, Hall and Sandler identified faculty behaviors that may account for the decline in academic and career aspirations many women experience in their college years, especially women in traditionally male majors.

Although individual instructors have an important part to play in encouraging women in mathematics, institutions also are important. The Task Force on Women, Minorities, and the Handicapped in Science and Technology (1989) outlines actions needed at all levels to increase the representation of women in the sciences. Even though more research is needed, we know enough of what works at the pre–college level (American Association for the Advancement of Science, 1984) to design college–level interventions.

Minority Issues. If the status of women in mathematics could be characterized as disturbing, the situation of most minority groups is alarming (Brown, 1987; National Science Foundation, 1988). African Americans represent only 2 percent of all employed scientists and engineers, and they earn only 4 percent of baccalaureates and 1 percent of the doctorates in science and engineering (Task Force on Women, Minorities, and the Handicapped in Science and Technology, 1988). Similar figures hold for Hispanics who, at 9 percent of our population, are the fastest–growing minority group.

There has been substantially less research focused on minorities and mathematics than on women and mathematics. Although previous mathematics achievement remains the best predictor of success in college, affective variables are important in predicting African Americans' success in college, particularly interests, motivation, and values (Atwater & Simpson, 1984). Encouragement to major in mathematics and participation in mathematics clubs were also positively related to interest in mathematics for African American males and females (Thomas, 1986).

Instructors at all levels need to be sensitive to linguistic and cultural variations in instruction which can promote learning. For bilingual students, there may be a threshold level of linguistic competence needed to avoid cognitive deficits (Mestre, Gerace & Lochhead, 1982). Relatively subtle aspects of interaction etiquettes are likely to go unrecognized by non–minority instructors, who may unwittingly cause offense or discouragement. Expert help from campus student services, and faculty development activities which focus on these issues, could help us learn to work with a diverse population.

A college–level intervention which has received considerable national attention is the mathematics workshop program at UC–Berkeley (Asera, 1988; Treisman, 1985). In contrast to most minority programs which are primarily remedial in nature, the Berkeley program offers an honors workshop in which students work individually and collaboratively on difficult mathematical problems. Many other campus services such as counseling, advising, and personal support are provided within this program. Fullilove and Treisman (1990) and Treisman (1985) report results for African American students [unfortunately not broken down by gender] which include: higher course grades; lower failure rate in first-semester calculus; higher retention and graduation rates; and higher rates of mathematics–based majors for workshop participants than for African American non–participants. Although the entering students at other campuses may be different from those at Berkeley, this intervention merits a close look by other departments to determine which elements are applicable in their settings.

SUMMARY. When discussing educational equity, there are various interpretations of the terms one might use. It is useful to consider, at the minimum, opportunity to learn, outcomes, and treatment as components of equity. While we may have legal equality of opportunity to learn, we cannot achieve equity, or justice, without equality of outcomes in mathematics education for all groups. That would include: equality of goals for both genders and all groups; no differences in what males and females, or African Americans, European Americans, Hispanics, Asians, and Native Americans have learned in mathematics; no differences in how any of these groups feel about themselves as learners of mathematics; and no differences in willingness to pursue mathematics and related careers. College faculty should acknowledge their role in attaining educational equity and accept the challenge which we all face in meeting that goal through further research and well–documented interventions.

ACADEMIC PREPARATION FOR COLLEGE

The number of college courses covering traditional high school courses, usually called developmental mathematics, has grown substantially in the last 30 years across the country. In this section we discuss what is known about these remedial approaches.

There is a clear pattern that, as colleges offered admittance to less well–prepared students and offered remedial courses, students took fewer mathematics courses in high school. Of course we claim no cause and effect here, and the change in colleges' policies were frequently made out of a sincere desire to provide opportunity to more students. However, the rapid growth in enrollment in developmental mathematics courses needs to be questioned. Early testing programs can have a positive effect on student enrollment and achievement (Adcock, Leitzel & Waits, 1981; Leitzel, 1983), and should be considered for wider use.

Regardless, we will continue to have students without adequate background entering the first college course. What should we offer such students? Greenes and Fitzgerald (1991) found that their remedial courses were not targeting students' difficulties nor preparing them for further coursework. Currently they are at work developing modules to target student difficulties identified by a computer–presented assessment.

Unfortunately, research provides little additional help in designing programs for remediation. Many remedial approaches have used some form of self–paced instruction; these approaches have met with mixed results and certainly cannot be generally recommended. MacGregor, Shapiro and Niemier (1989) did find that computer–augmented learning was beneficial for field dependent students in developmental algebra. However, those with indiscriminate cognitive styles demonstrated greater achievement with traditional instruction. Perhaps more aptitude–treatment interaction studies are needed to help develop useful methods for remedial students with various cognitive styles.

There is some indication that test anxiety is an important factor in the mathematical achievement of remedial students, and in fact may be more important than mathematics anxiety (Green, 1990). Green also found that teachers' written comments on student work facilitated achievement.

A critical question to ask about remedial programs is how well students do in regular coursework taken subsequently. In many cases, the answer is, not well at all. For example, Eisenberg (1981) followed 1600 students who had taken beginning algebra at Northern Michigan University for four years, examining their performance in following courses. Sixty percent of the A/B students did not take another mathematics class, with higher percentages of C and lower grades opting out of further mathematics. Only 35 percent of the algebra students as a whole took another mathematics class; of these, 45 percent failed.

TEACHER EDUCATION

Although we think this is a critical area for research, the paucity of published research on mathematics-teacher education leaves us with little to write in this section. We would like the reader to be clear that we are making a distinction between research on teaching and research on teacher education. Most of the research has focused on ascertaining teachers' [pre– and inservice] knowledge in and attitudes toward mathematics, with little investigation of the effect of different models of teacher education on classroom behavior. Very few studies have been done of secondary teachers, whom most mathematics departments have the major role in educating.

Research on teacher education needs to focus on content knowledge, teacher attitudes and beliefs, and pedagogy. Some of the recent work on novice teachers (Cooney, 1985) or comparisons between novice and expert teachers (Borko & Livingston, 1989; Livingston & Borko, 1990) point out that our teacher education programs cannot merely present student teachers with information about how expert teachers teach and expect it to be emulated. Research is needed to help determine what else is required.

Much more information is needed about the process through which teachers change. Several ongoing research studies hold the promise of providing results critical to implementing reform in the teaching of mathematics (Hart, 1991; Stein, Grover, & Silver, 1991). Teacher beliefs are also a critical variable to consider in preservice or inservice programs (Becker & Pence, 1990; Cooney, 1985; Dougherty, 1990; Jurdak, 1991; Russell & Corwin, 1991). It may not be possible to change teacher behavior without a change in teacher beliefs about mathematics and about the learning and teaching of mathematics. One approach to improvement of instruction--peer coaching--was found successful in a study by Tobin and Espinet (1990); peer coaching merits further investigation.

In light of the sweeping changes in instruction called for by the National Council of Teachers of Mathematics (NCTM, 1989) and other organizations, the efficacy of models of mathematics teacher education needs investigation. The state of research on mathematics teacher education is not healthy, and would benefit from a sustained focus from mathematics teacher educators.

FUTURE DIRECTIONS

As the review of existing research shows, there is a growing body of knowledge on the teaching and learning of mathematics at the college level. However, many questions remain to be answered.

Curriculum research should focus not only on evaluating and improving specific materials but also on assessing the qualitative differences among different approaches. Research on student learning needs to go into the classroom to see how learning occurs in context. New models are needed which relate teaching and learning.

At least two types of research are needed on pedagogy. We need to determine how best to help mathematics faculty develop expertise in new teaching methods. And we need research to determine the impact of new methods and how best to implement them, particularly for students currently underrepresented in mathematics.

We need research to help us understand the social context of learning at the college level, and the linkage between cognition and affect. Research on equity issues should be an important component of any intervention programs. And finally, mathematics-teacher education needs to become a topic for systematic investigation.

Overall, research should work toward developing theoretical models which can guide subsequent studies. This approach will be more likely to result in a coherent body of usable knowledge on a particular question than individual, uncoordinated studies such as much of the previous research.

PART II
OBSERVATION/ASSESSMENT

A sample of A and B students from a very standard calculus course showed that their competence in calculus is restricted to the solution of problems that are well rehearsed. There is every reason to believe that this result applies to hundreds of thousands of students in the United States. Another factor apparent from this study is that students will tend to use primitive methods in situations where most mathematicians would regard calculus methods as the most appropriate. Analogous phenomena have been observed among high-school algebra students, who tend to use numerical methods to solve problems that most would regard as algebra problems; e.g., situations that we would regard as being appropriately modeled by simultaneous equations are approached numerically in trial and error fashion.

One question is: what are we to make of this? On one hand, this sort of dipstick into the status quo provides superficial information about the shallow impact of our instruction and our curriculum, but it may also provide more helpful information if we look more closely. In the case of the simultaneous equations example, we might infer that numerically based approaches provide a natural route into the target topic. Simultaneous equations come to be seen as a more efficient method for doing something that was previously established as quantitatively meaningful because it is soundly rooted in prior arithmetic experience in satisfying simultaneous quantitative constraints, rather than as a meaningless game played with symbols.

So now, what analogous lesson might be learned from the calculus study? The answer may not be as simple, partly because the topics are much more varied in this study. On one hand, there certainly are many instances where prior quantitative experience, perhaps using algebraic techniques in some cases, provides the base for calculus learning. And one might argue that these, rather than the usual technique–oriented courses, deserve the role of "precalculus" courses. But, if we look at the tasks used in more detail, deeper questions arise.

Three of the five non–routine problems involve parametrizations. In fact, two of them require coordinating two parametrized conditions, each of which defines a set of functions. The authors describe these as requiring "the students to combine what should have been familiar techniques and concepts in a way new to them, but the solutions are no more complex than those of many sample problems given during the course." On one hand, we agree that the complexity in terms of visible problem–solving steps is comparable to standard course problems. However, the conceptual complexity of the mental objects involved is decidedly more complex. It is one thing to deal with the differentiability of a piecewise defined function (which most students, and most mathematicians up to the 19th century, would regard as multiple functions), and it is entirely another to coordinate two parametrized classes of functions over adjacent intervals to achieve differentiability at the boundary.

The rather large-scale failure rates on these and like problems seems to indicate that the underlying concepts of continuity and differentiability are not seen as applying to the elements of these classes. Indeed, we suspect (and it is only a suspicion) that the students do not even conceptualize functions in these situations, but rather see only ill–defined expressions involving extra unknowns whose role is unclear. For them the phrases "the line $2x + 3y = a$" and "the graph of $f(x) = bx^2$" are likely fraught with mystery. Use of the singular article "the" to denote what they are likely to regard as a multiplicity must be confusing, unless they have conceptualized the set being referred to as a unity. Since they are unlikely to have had systematic opportunity to develop such conceptualizations, failure seems predictable.

So at a cognitive level, this study helps point to rather specific but significant shortcomings in the curriculum--failures to provide opportunity to build conceptualizations of parametrized classes of functions. Without these conceptualizations, any basic ideas that the students may have about properties of functions probably cannot be applied. It remains an open question whether the test of routine problems, all referring to single functions, actually tapped these basic ideas, and whether these ideas were really present in the minds of the students. And there were some rather disturbing data from the routine problem test: for example, a majority of A and B students, being told that 5 is a root of $f(x) = 0$, could not tell where the graph of $y = f(x)$ crosses the x-axis. The reader is invited to examine the correspondence between routine and non–routine problems offered by the authors and ask what is the essential new ingredient required to do the corresponding non–routine problem.

Many interesting research questions are indirectly raised by this paper. But, since the authors did such a careful job of showing how representative of the typical calculus student these students are, the direct message of the study is also highly relevant to reform considerations. The direct message should be quite disheartening for those who believe in the current curriculum.

Even Good Calculus Students Can't Solve Nonroutine Problems

John Selden, Annie Selden, and Alice Mason
Tennessee Technological University

Abstract

In a previous study we found that C students from our traditional first-quarter calculus course could not solve cognitively nontrivial problems, that is, problems for which they had not been taught a method of solution. The present paper reports on a similar study of our traditionally taught A and B calculus students. Although they performed slightly better on our test of nonroutine problems, two–thirds of the students failed to solve a single problem completely and more than 40% did not make substantial progress on a single problem. A subsequent routine test confirmed that these students possessed an adequate knowledge base of relevant calculus skills. This suggests that traditional methods of teaching calculus are insufficient in preparing even good students to apply calculus creatively. Furthermore, students from both studies exhibited a large tendency not to use calculus, preferring more elementary solution methods.

Introduction

A previous study (Selden, Mason, and Selden, 1989) confirmed the extremely limited problem–solving abilities of C students from traditional calculus courses and raised the question of whether A and B students from such courses would perform better when confronted with nonroutine or novel problems. It also raised the issue of whether the students' routine calculus skills were inadequate or whether they lacked some additional ability. In the present paper, we report on a study[1] which examines these questions.

A *problem* in the sense used here can be regarded as having two components, a task and a solver. Solving the problem consists of finding a correct method of solution and carrying it out. Although perhaps not explicitly mentioned in some problem–solving studies (Schoenfeld, 1985), the solver, who comes equipped with (possibly imperfect) information and skills, is an essential part of this view of a problem. Such studies assume the tasks are *cognitively nontrivial,* that is, that the solver does not begin already knowing a method of solution. This means that, in a certain sense, problems cannot be solved twice by the same individual. Of course, the solver may forget the solution, but the degree to which the solution has been forgotten is very difficult to assess. Thus tasks, in themselves, cannot be classified as problems independent of knowledge of the solver's background. Traditional calculus courses contain few such cognitively nontrivial problems, preferring to include tasks which should more appropriately be termed exercises. However, even tasks only slightly different from examples in calculus texts can become problems in our sense, as was seen in the nonroutine test we prepared for our C students (Selden, et al, 1989, p. 48).

In gauging the calculus problem–solving ability of our A and B students we employed the same test previously administered to C students. We also gave a second test whose questions covered precisely those routine skills we judged were needed for solving the nonroutine problems on the first test. The second test was for the purpose of distinguishing between a lack of routine skills and a lack of some other ability. What follows is a description of the course, the students, the tests, and the results.

The Course

Tennessee Technological University is a comprehensive state university, with an engineering emphasis, enrolling about 7500 students. The average ACT composite score of the 1,461 entering Freshman in Fall 1988 was 20, which is above average for the state.

The pool of students considered came from all sections of the first quarter of our mainstream calculus, omitting our one experimental and our one honors sections. This course serves all mathematics, science, and

---
[1] Supported by The Mathematics Department of Tennessee Technological University.

engineering students desiring to take calculus. A separate course is offered for business majors and pre-professional students. Classes for the course met 5 hours per week for 10 weeks in Fall Quarter 1988 and were taught from *Dennis P. Berkey, Calculus, Second Edition*, Chapters 1–4, plus coverage of mathematical induction and complex numbers from Appendices II and III (a routine change of text from the previous year). The 8 nonhonors, nonexperimental sections, consisting of about 38 students each, were taught by 6 different instructors. These were taught by regular full–time faculty members of all ranks. All instructors taught according to their own normal methods and handled their own examinations and grades. None of this was unusual in any way. The engineering students, the bulk of the clientele in this course, had taken the MAA Calculus Readiness Test and had been advised accordingly. Additionally, the catalog states that a math ACT score of at least 26 is required for direct entry into the course, although this is not always enforced.

In our pool, 62.3% of all students and 78.8% of the A and B students were taking calculus as their first college mathematics course. Overall, there were 10.5% A's, 17.4% B's, 20.1% C's, 10.5% D's, 33.2% F's, and 8.2% W's. This is a smaller percentage of A's and B's than the previous year when we studied the C students (27.9% vs. 45.1%). However, that study included our one section of honors students, most of whom earned A or B. In addition, there is considerable variability in the grading styles of instructors (Flener, 1990); only one instructor taught our mainstream first quarter calculus both years.

THE STUDENTS

At the beginning of Spring Quarter, the 85 students who completed the first quarter of calculus with A or B (32 A and 53 B) were contacted by mail and invited to participate in the study. Each student was offered a fee of $15 for taking the tests and told he or she need not, in fact should not, study for them. As an additional incentive to ensure all students would be motivated to do their best, the A students and the B students were separately offered prizes of $20, $15, $10 and $5 for the top scores. They were told nothing about the test except that it would involve first-quarter calculus.

There were 20 volunteers from the A students and 19 volunteers from the B students. We randomly selected 10 from each group. This volunteer rate of 46% is only slightly higher than the 40% volunteer rate of the previous year's C students. Of those selected, 10 A and 9 B students actually took the test; 16 of the 19 were engineering majors. Five of the six instructors of the nonhonors, nonexperimental sections were represented by the randomly selected students.

Our random selection divided the volunteers into two groups with very similar characteristics with respect to mean ACT, mean math ACT, GPA (on a 4 point scale) at the end of Winter Quarter, and Calculus II grades for those who continued. In Table 1 these are indicated by the headings ACT, MATH ACT, GPA, and A, B, C, D, F, W, respectively. Eight students did not continue calculus in Winter Quarter; of these one volunteered and was selected for the study. Students who did not volunteer for the study had mean ACT and mean math ACT scores similar to those of the volunteers, but somewhat lower mean GPA and Calculus II grades. Thus our sample students could be expected to perform at least as well as the entire pool of A and B students, had we tested them all.

Table 1.
Characteristics of the A and B Students

	Volunteers Tested (19)		Volunteers Not Tested (20)		Non–Volunteers (46)		Total (85)	
ACT	27.12		26.55		26.59		26.70	
Math ACT	28.00		29.60		27.92		28.38	
GPA (max 4.0)	3.264		3.263		3.028		3.139	
A	7	(38.9%)	9	(45.0%)	5	(12.8%)	21	(27.3%)
B	5	(27.7%)	4	(20.0%)	14	(35.9%)	23	(29.9%)
C	5	(27.7%)	4	(20.0%)	9	(23.1%)	18	(23.4%)
D	1	(5.6%)	0	(0.0%)	6	(15.4%)	8	(10.4%)
F	0	(0.0%)	0	(0.0%)	3	(7.7%)	3	(3.9%)
W	0	(0.0%)	3	(15.0%)	2	(5.1%)	4	(5.2%)
Total Continuing	18		20		39		77	

THE TESTS

Two tests were given, one immediately after the other. The first test lasted one hour and consisted of the five nonroutine problems given to C students in our previous study. The second test lasted half an hour and consisted of ten routine questions, covering precisely those skills we judged were needed to solve the problems on the first test.

The nonroutine problems required the students to combine what should have been familiar techniques and concepts in a way new to them, but the solutions are no more complex than those of many sample problems given during the course. In order to determine whether the students had an adequate knowledge base for solving the nonroutine problems, we chose *very* simple, straightforward routine questions that would minimize the occurrence of computational errors.

NONROUTINE TEST

1. Find values of a and b so that the line $2x+3y=a$ is tangent to the graph of $f(x)=bx^2$ at the point where $x=3$.

2. Does $x^{21}+x^{19}+x^{-1}+2=0$ have any roots between -1 and 0? Why or why not?

3. Let $f(x)=\begin{cases} ax, & x\le 1 \\ bx^2+x+1, & x>1 \end{cases}$. Find a and b so that f is differentiable at 1.

4. Find at least one solution to the equation $4x^3-x^4=30$ or explain why no such solution exists.

5. Is there an a so that $\lim\limits_{x\to 3}\dfrac{2x^2-3ax+x-a-1}{x^2-2x-3}$ exists? Explain your answer.

Due to blurred photocopying, x^{19} in Problem 2 was read by all students in the present study as x^{10}; consequently, all were attempting to solve an accidentally altered problem. The altered problem is somewhat more difficult than the original for solutions using the derivative, but comparable for solutions from first principles. Only one student in this study attempted a solution method which might have been more successful on the original problem.

ROUTINE TEST

1. (a) What is the slope of the line tangent to $y=x^2$ at $x=1$?

   (b) At what point does that tangent line touch the graph of $y=x^2$?

2. Find the slope of the line $x+3y=5$.

3. If $f(x)=x^5+x$, where is f increasing?

4. If $f(x)=x^{-1}$, find $f'(x)$.

5. (a) Suppose f is a differentiable function. Does f have to be continuous?

   (b) Is $f(x)=\begin{cases} x, & x>0 \\ 2, & x\le 0 \end{cases}$ continuous?

6. Find the maximum value of $f(x)=-2+2x-x^2$.

7. Find $\lim\limits_{x\to 1}\dfrac{x^2-1}{x-1}$

8. Do the indicated division: $x-1\overline{)x^3-x^2+x-1}$.

9. If 5 is a root of $f(x)=0$, at what point (if any) does the graph of $y=f(x)$ cross the x–axis?

10. Consider $f(x)=\begin{cases} x^2, & x\le 1 \\ x+3, & x>1 \end{cases}$.

    (a) Find $\lim\limits_{x\to 1^+} f(x)$

(b)   What is the derivative of $f(x)$ from the left at $x = 1$ (sometimes called the left–hand derivative)?

Before the Nonroutine Test commenced, students were cautioned they might find some of the problems a bit unusual. They were asked to write down as many of their ideas as possible because this would be helpful to us and to their advantage. They were told A students would only be competing against other A students for prizes and, similarly, B students would only be competing against other B students. They were assured all prizes would be awarded and that partial credit would be given.

Each nonroutine problem was printed on a separate page on which all work was to be done. All students appeared to be working diligently for the entire hour. As in the previous study, each problem was assigned 20 points and was graded by one of the authors and checked by the others. Grading was reasonably liberal when compared to that of most calculus teachers or the Advanced Placement Calculus Test. The highest score was 67 out of 100, obtained by a B student. The highest score in our previous study of C students was 35. The lowest score for both studies was 0. The mean test score was 20.4 for the A and B students, as compared to a mean of 10.2 for the C students.

As soon as the Nonroutine Test was collected, the students were given the Routine Test, which was on two pages. They were told answers without explanations would be acceptable, but they could show their work if they wished. Most students worked quickly; a few completed their work and were restless after 20 minutes. When one student requested permission to leave and was allowed to do so, others followed rapidly. No one stayed the allotted amount of time. Each question was assigned 10 points and was graded by one of the authors and checked by the others. The highest score was 90 out of 100; the lowest score was 53. The mean score was 73.4. See Table 2 for the distribution of scores.

Table 2.
Test Scores

Nonroutine Score	Routine Score	Calculus I Grade
67	86	B
47	88	A
41	88	A
41	72	B
26	83	A
24	83	A
24	65	B
21	85	A
21	55	B
18	61	B
16	90	A
12	83	B
9	53	B
8	64	A
6	60	A
3	80	A
2	60	A
1	85	B
0	53	B

THE RESULTS

About two–thirds of the students (7 A and 5 B) could not get a single problem correct on the Nonroutine Test. While not as extreme a result as that for C students in our previous study, where not a single problem was solved correctly, it is still a remarkable result for our best students.

We will refer to each page of written work submitted in answer to the Nonroutine Test as a solution *attempt*, making a total of 95 (5 times 19) solution attempts. A number of solution attempts contained several distinct approaches to solving the problem — each of these is referred to as a solution *try*. On examining the 95 solution attempts, we found a total of 113 solution tries to the five nonroutine problems, not counting the 17 crossed–out solution tries to Problem 1 given by one student. Of these, 9 were judged *completely correct*. Nine others were judged *substantially correct* because they exhibited substantial progress towards a correct solution, meaning the proposed solution could have been altered or completed to arrive at a correct solution and was assigned at least 10 points. The 9 correct solutions were submitted by 7 students with Problem 1 solved three times, Problem 4 solved once, and Problem 5 solved five times. The 9 substantially correct solutions were submitted by 6 students with Problem 1 solved four times, Problem 2 solved once, Problem 3 solved once, Problem 4 solved once, and Problem 5 solved twice.

Every instructor represented in the study had at least one student get one nonroutine problem correct or substantially correct, and we detected no bias in the results toward a particular instructor. Altogether 11 students gave correct or partially correct solutions, which leaves 42% (8 out of 19) of the A and B students *unable* to make substantial progress on *any* nonroutine problem. Again, this is less extreme than the 71% of C students (12 out of 17) who could not make substantial progress on these problems in the previous study, but it is alarming that such a large proportion of our best students are unable to use calculus in a flexible manner.

Of the 76 solution attempts to the first four problems, 46 made *no* use of calculus whatever; 11 made only perfunctory use of calculus, for example, taking a derivative and ignoring it in the remaining work. Converting these numbers to percentages and comparing them with our previous study of C students, we were surprised to find that essentially the same percentage of students in both studies used no calculus whatever (61% vs. 59%) or essentially no calculus (75% for both) in an attempt to solve problems half of which contained the key words "tangent" or "differentiable." The vast majority of the students in both studies had continued calculus in Winter Quarter (95% vs. 88%) and thus should have been current in their use of it. It is noteworthy that earning a better grade in first-quarter calculus does not seem to imply that one will be more inclined to use it.

The routine questions were designed to check whether students' inability to do the nonroutine problems related to an inadequate knowledge base. Did the students have the necessary factual knowledge without being able to access it? Scores on corresponding routine questions (Table 3) were taken as indicating the extent of students' factual knowledge regarding a particular nonroutine problem. A student was considered to have *substantial factual knowledge* for solving a nonroutine problem if that student scored more than 66% on the corresponding routine questions. A student was considered to have *full factual knowledge* for solving a nonroutine problem if that student's answers to the corresponding routine questions were correct, except possibly for notation, for example, answering $(1,-1)$ instead of -1 to Question 6.

Table 3.
Correspondence of Routine Questions with Nonroutine Problems

Nonroutine Problem	1	2	3	4	5
Corresponding Routine Questions	1,2	3,4,9	5,10	6,9	7,8

We examined the relationship between students' possession of substantial factual knowledge (62 attempts) and their performance on corresponding problems of the Nonroutine Test. This was done separately for full factual knowledge and substantial, but not full, factual knowledge. The results are summarized in Table 4 and suggest that, while factual knowledge is important for the solution of nonroutine problems, most students cannot access it effectively. We note, however, that the relationship between possession of substantial factual knowledge and the ability to solve nonroutine problems may be quite complex. Some studies suggest that misconceptions, when present, are more likely to surface during nonroutine problem solution, even though students are able to answer routine questions correctly (Amit and Vinner, p. 3; Eylon and Lynn, p. 257). Furthermore, it is possible that the careful analysis of a nonroutine problem might actually improve access to factual knowledge. In our study, two B students with *incomplete factual knowledge*, as evidenced by scores of 40% and 50% on the corresponding routine questions, each submitted one completely correct solution, one to Problem 4 and one to Problem 5.

Table 4
Nonroutine Problems Solved and Factual Knowledge

Factual Knowledge (attempts)	Substantially or Completely Correct		Completely Correct	
Substantial, but not full (23)	13%	(3)	0%	(0)
Full (39)	33%	(13)	18%	(7)

Periodic monitoring and assessment of solutions is one major component of effective control in problem solving (Schoenfeld, p. 33), and is one characteristic of expert problem-solvers. We analyzed the students' use of monitoring as indicated by scratchwork, abandoned methods of solution, and checking of answers, and we found considerable evidence of its use in 36% (34 out of 95) of the solution attempts. However, in just 22 of these attempts, students also possessed substantial or full factual knowledge (Table 5).

Table 5
Students who Monitored their Work
Nonroutine Problems Solved and Factual Knowledge

Factual Knowledge (attempts)	Substantially or Completely Correct		Completely Correct	
Substantial, but not full (5)	20%	(1)	0%	(0)
Full (17)	47%	(8)	24%	(4)

Comparing Tables 4 and 5, we see that monitoring in the presence of substantial factual knowledge helps, but does not guarantee success. Other aspects of control are also worth considering. For example, the student who made 17 tries on Problem 1 and a total of 25 tries in all on the Nonroutine Test exhibited considerable monitoring and much activity, but little effective control. Not surprisingly, that student was the only one who scored 0 on the Nonroutine Test.

We also considered the use of graphs by our A and B students in working both tests. Graphs were used in only 26% (20 out of 76) of the solution attempts on Problems 1 through 4 of the Nonroutine Test. Problem 5 was judged less amenable to graphical representation and hence, was not included in this analysis. Graphs were drawn in 11 attempts on Problem 1, 6 attempts on Problem 3, and 3 attempts on Problem 4. Students only sometimes used their graphs effectively, and in one instance, a graph led the student astray. Furthermore, on Question 1 (b) of the Routine Test 53% of the students (10 of 19) could not correctly indicate at what point the tangent line touches the graph of $y = x^2$ given $x = 1$. In answering Question 9, 58% (11 out of 19) could not correctly state at what point the graph of $y = f(x)$ crosses the x–axis, given 5 is a root of $f(x) = 0$. The only question with more incorrect responses (12 out of 19) than Question 9 was Question 10(b) on left–hand derivatives; however, three–quarters of the incorrect responses (9 out of 12) were "$2x$", for which partial credit was given. Our observations agree with those of others that the graphical knowledge of traditional calculus students is weaker than their analytic knowledge (Dreyfus, p. 12; Dreyfus and Eisenberg, p. 27; Dick, p. 2) and should be improved as, indeed, a number of current calculus reform projects are attempting. They also indicate why the scores on the Nonroutine Test were lower than we expected.

One engineering student in this study, whom we will call J.W., illustrates our findings well. J.W. arrived at Tennessee Tech in the Fall of 1988 with an ACT composite score of 25 and a math ACT score of 27. J.W. had studied calculus in high school and was granted credit for first-quarter calculus, based on an Advanced Placement AB Test score of 4, but took our course anyway, earning an A. In Winter Quarter, J.W. continued with calculus, again earning an A, and had an overall GPA of 3.514 at the end of that quarter. There was considerable variation in the proportion of high grades given by our calculus teachers, but both of J.W.'s teachers were close to the median in this regard. They describe J.W. as a quiet, unassuming student who asked few questions but was always present. By all the usual standards, J.W. was an excellent student, well–versed in first quarter calculus. J.W. made the highest, and only, grade of 90 on our Routine Test, but scored just 16 on the Nonroutine Test, getting only Problem 1 substantially correct.

FAVORED SOLUTION METHODS

On Problem 1 the favored method of solution, given by 10 students, and also favored by the C students in our previous study, was to solve the equations of the line and the parabola simultaneously, thereby not using the crucial information that the line was tangent to the parabola. The next most favored method was the correct one, employed by seven students, representing 4 of the 5 different instructors. Six students did not take a derivative, and three others took the derivative but made essentially no use of it. Four students calculated the slope of the parabola using two points on the parabola and the "rise over run" formula, thereby confusing "average slope" with "slope" — a conceptual error.

On Problem 2 there were three equally favored methods of solution, each seen in 5 attempts. The most effective of these, which led to one substantially correct solution and partial credit in other cases, was a first principles argument based upon comparing the relative sizes of x^{21}, x^{10}, and x^{-1} for numbers ranging between -1 and 0. Another method was to factor $x^{21} + x^{10} + x^{-1}$ in some way and either give up or set both factors equal to -2. The final method was simply to use trial and error, that is, to substitute a few numbers for x and calculate the size of $x^{21} + x^{10} + x^{-1} + 2$ for these; this was the favored method of the C students in our previous study. Two students attempted to use the Theorem on the Rational Roots of a Polynomial, and one student tried to use Descartes' Rule of Signs, indicating each has a substantial high-school algebra course, which came to mind more readily than their more recent calculus. Fourteen students made no use whatsoever of calculus, preferring more elementary methods, and *none* used calculus effectively.

On Problem 3 seven students set $ax = bx^2 + x + 1$ and substituted $x = 1$ to get $a = b + 2$, but could go no further. This method was also favored by the C students. Thirteen students did not take a derivative. Five students tried to take a derivative (of whom two calculated the derivative of the function as the *sum* of the left–hand derivative and the right–hand derivative!); however, none was able to make effective use of it. Five tried to use the concept of differentiability, with one stating correctly that differentiability implies continuity and one stating the converse. Four drew good graphical representations of the problem joining a parabola to a straight line, but only the one student who solved the problem substantially correctly seemed to gain usable insight from the graph.

On Problem 4, six students took a derivative to find critical numbers or to consider where the function is increasing or decreasing. Five students employed first-principles arguments, exploring intervals where $4x^3 - x^4$ is negative or positive. Only one student substituted a few values for x in $4x^3 - x^4$ and guessed-- the favorite method of the C students. Four students tried to reduce the equation to a quadratic in some way and use the quadratic formula. Altogether there were 25 solution tries, only six of which used any calculus. A number of tries exhibited the use of fairly sophisticated results from high school algebra: Descartes' Rule of Signs (2 tries), the Theorem on Rational Roots of a Polynomial (4 tries), and synthetic division (3 tries). Thus, it appears that our A and B student had significantly more algebraic skills and solution strategies at their command than did our C students. They also tended to explore what happened for general values of x, rather than substituting specific numbers. On the other hand, four of them factored $4x^3 - x^4$ and set each factor equal to 30, a favored method of the C students. One of these students also used the Theorem on Rational Roots of a Polynomial correctly, as well as setting the derivative of $4x^3 - x^4$ equal to 0 to look for extrema, thereby illustrating that good technical knowledge and misconceptions can coexist.

On Problem 5, seven students representing four of the five different instructors, obtained correct or substantially correct answers. This was the only problem in which correct methods were the favored ones. However, it was also the only one whose solution requires principally algebraic techniques. Four of the seven students set the numerator equal to 0, substituted $x = 3$, and solved for a. Two noted that $x - 3$ must be a factor of the numerator. One of these divided the numerator by $x - 3$, set the remainder equal to 0, and solved for a; the other factored the numerator into a product of $x - 3$ and a linear factor involving a, multiplied these out, and matched the resulting middle term involving a with that of the original numerator. The remaining student substituted $x - 3$ in the expression, obtained $a = 2$ by inspection, and checked to see whether this was correct. Two of the seven used L'Hôpital's Rule to check their answers. Five students (26%) substituted $x = 3$ in the function, found the denominator was 0, and concluded the limit could not exist; this was the favored method of the C students (47%). Four students divided the numerator and denominator by x^2, an appropriate technique when $x \to \infty$.

CONCLUSIONS AND QUESTIONS

On the whole, good students from traditionally taught calculus classes cannot be expected to solve many nonroutine problems. In this study, two–thirds of the students could not solve a single problem correctly, and many (42%) did not show substantial progress on any problem. Furthermore, these results were not due principally to the students' lack of factual knowledge. Only one–third of the solution attempts in the presence of full factual knowledge showed substantial progress, and only about one–fifth (18%) were completely correct. Students who also monitored their work fared somewhat better. Of the solution attempts showing monitoring and full factual knowledge, one–fourth were correct, and about half (47%) showed substantial progress.

Many students in this study preferred arithmetic and even quite sophisticated algebraic techniques for solving calculus problems. They seemed to have better access to these elementary techniques, but were often unable to recognize situations in which their use is inappropriate. Over half of the students in both our studies made no use whatever of calculus. Perhaps it takes time to feel comfortable with a new subject. Would students at the end of our two–year calculus–differential equations sequence be more inclined to use calculus techniques in solving problems?

Graphs were used in only one–fourth of the solution attempts to the first four nonroutine problems. As others have observed, proper interpretation of graphs appears to be quite difficult.

Since not all of the diverse applications of calculus can be made routine, our results suggest the usefulness of developing teaching techniques that improve students' ability to solve nonroutine problems. In this regard, the mere accumulation of additional factual knowledge is unlikely to improve nonroutine problem–solving ability. In our discussion of students' favored solution methods, we noted a number of conceptual weaknesses. It would be interesting to investigate the relative contributions to nonroutine problem-solution of conceptual grasp versus general problem–solving techniques such as heuristics and control.

Finally, our results suggest calculus reform needs to concentrate not only on increasing the success rate of all students, but also on improving their ability to apply calculus creatively. To accomplish this, the teaching of calculus might well need to be structured differently.

PART III
INTERPRETATION

Using extended interviews with a small number of students, these authors examine these students' understanding in a small number of students' understanding, of the central ideas of calculus, function, limit, continuity, derivative, and integral. They report on one student in detail, who reveals a startling contrast between ability to use algorithmic approaches to routine problems and underlying understandings of the concepts involved. While the particulars of this contrast may vary from student to student, there is good reason to believe that the disturbing lack of conceptual foundations for this student's procedural competence may be widespread. The student and her instruction appear to be entirely representative of the norm in the United States.

The superficiality of her understanding parallels that found in the Seldon–Mason studies, but the parallels reach farther, especially in the gap between students' understandings of symbolically based representations of the basic concepts and the graphical realization of those concepts. These appear to be entirely unrelated to one another. A student who can, with considerable facility, apply differentiation and integration rules to functions described symbolically, frequently cannot explain or even discuss the graphical interpretation of these actions. And, even when the student was able to carry out the typical graphing activities of a calculus course using derivatives in the usual way, she was unable to describe how the tangent to the graph of a function is related to the derivative of that function. This is reminiscent of the student in the Seldon–Mason study who, given information about the roots of a function, could not tell where its graph crossed the x–axis. So not only is the knowledge superficial, it is extremely fractured. If this were not enough, the authors expose underlying student beliefs about the nature of mathematics that make the harboring of inconsistent ideas acceptable and natural, and that leave to the external authority of texts and instructors the responsibility for determining the adequacy of mathematical reasoning or its results.

This gloomy story is, of course, the research–based part of the rationale for the calculus reform movement. While there is reason to hope for improvement based on new curricula, new pedagogies, and increased application of technologies that increase engagement and involve linked representations of important ideas, we see strong need to move beyond the unsettling results of this sort of study towards both more detailed and more comprehensive analyses of the structure of student knowledge of calculus and related concepts.

However, mapping out student knowledge that is a result of a curriculum and forms of practice that we already acknowledge as seriously deficient may not be the most efficient way to proceed. More efficient, perhaps, may be an approach that measures the impact of instruction that more closely approximates our current ideas of what may be more effective. Such instruction may also need to assume a student background substantially different from that of students arriving on campus today. Nonetheless, the kinds of results described in this chapter provide a baseline but, even more importantly, the qualitative and descriptive methods can be applied to great advantage, especially if augmented by an historical perspective. Note, for example, that the very widespread, but limited, student conceptions of function and continuity are reasonably representative of the evolving notions in the 18th century. It might prove instructive to determine the factors and incidents that advanced the historical understandings of these ideas in earlier times.

RESEARCH IN CALCULUS LEARNING: UNDERSTANDING OF LIMITS, DERIVATIVES, AND INTEGRALS

Joan Ferrini–Mundy and Karen Graham
University of New Hampshire

INTRODUCTION

As professionals in mathematics education, we are fortunate to be participants in this era of dramatic reform in mathematics curriculum and instruction which is affecting grades K–14. Changes in the discipline of mathematics, changes in the demands of the workplace and the characteristics of the workforce, availability of technology, and new knowledge about mathematics -learning help form the rationale for documents such as the NCTM *Curriculum and Evaluation Standards for School Mathematics* (NCTM, 1989.)

In the case of calculus reform, a combination of factors has prompted the college mathematics-teaching community to be concerned with creating a "lean and lively" (Douglas, 1986) calculus experience that will "prime the pump" (Tucker, 1990.) Resulting experiments in curriculum development and instructional change promise to provide new and exciting possibilities for helping students to learn calculus more effectively. At the same time, there is continued interest within the mathematics-education research community in learning issues related to calculus and precalculus. The body of available research in these areas has been reviewed (Ferrini–Mundy & Lauten, 1993; Leinhardt, Zaslavsky, & Stein, 1990; Becker & Pence, this volume), and certain patterns of findings are emerging that may be useful in the reform process. The context of curricular reform provides new opportunities for the interaction of mathematics-education researchers interested in calculus learning, and mathematicians interested in curriculum change (Ferrini–Mundy & Graham, 1993.)

Shavelson (1988) cautions that a faulty assumption about educational research is that "education research should directly and immediately apply to a particular issue, problem, or decision" (p. 5). Silver (1990) offers the argument that several different aspects of research have the potential to contribute to educational practice, including research findings, research methods, and the theoretical perspectives that frame research. With regard to work in precalculus and calculus learning, it seems that Silver's argument fits especially well. Certainly the various interpretations of a constructivist perspective (von Glasersfeld, 1984) that guide much of the ongoing work in calculus, the interview and assessment tasks that are employed in the studies, and, to some degree, the patterns that are emerging through a series of studies, all have the potential to contribute to the practice of calculus curriculum development and instructional innovation.

This paper will present parts of a study intended to describe college calculus students' understandings of certain central ideas in calculus. In particular, through a series of clinical interviews we have identified features of student understanding of the notions of function, limit, continuity, derivative, and the definite integral. Preliminary description and analysis are presented here, and issues for the practice of calculus curriculum development and calculus instruction are discussed.

RESEARCH FRAMEWORK

THEORETICAL PERSPECTIVE

Our study is guided by the basic Piagetian constructivist view that "mathematics learning is a process in which students reorganize their activity to resolve situations that they find problematic." (Cobb, Wood, Yackel, Nicholls, Wheatley, Trigatti, & Perlwitz, 1991). Noddings (1990) offers several points which are generally agreed upon in a constructivist perspective, including the view that mathematical knowledge is constructed, at least in part, through a process of reflective abstraction, and that cognitive structures are under continual development. To hold such a perspective in the conduct of research implies certain basic assumptions. We assume that the students are making their own sense of the tasks and contexts provided, based on experience. We also must assume that the students' constructions are rational and subject to explanation. We view the student's constructions not as errors or misconceptions to be eradicated and replaced with the "correct" and publicly shared interpretations of major ideas, but rather as expected phenomena that are natural in the learning process. Thus the theoretical perspective guides the methodology and analysis of the data. For an interesting discussion of constructivist views in mathematics teaching and learning, see Davis, Maher, and Noddings (1990.)

Methodologies employed in studies of this type are often qualitative and descriptive, based on interviews with students as they complete mathematical tasks. Sample sizes are small. The intention is to provide rich and defensible descriptions of student understandings that can serve as springboards for acknowledging the great complexities to be understood in learning about student knowledge. Davis (1990) asks: what is it that science gives us? "Sometimes it is, indeed, some kind of generalization that can be coded in some symbolic or abstract form." He goes on to argue that the contributions of science might also be described as suggestions of kinds of things that "one ought, perhaps, to look for." He claims "those who disregard descriptions are seeing 'science' mainly as a collection of some kind of abstract generalizations, whereas an equally important part of science — perhaps a more important part — is the collection of metaphors it gives that allow us to think about the world in certain ways." See also the section in Schoenfeld, Smith, and Arcavi (in press) titled "On generality: How seriously can you take a study with $n=1$?" for an interesting discussion of the issue.

RESEARCH IN CALCULUS LEARNING

Emerging patterns of findings and hypotheses generated by the work of others in calculus learning served to guide and undergird the questions pursued in this study. The research available on understanding of function is quite substantial. Davis (1984) observed that students come to calculus with a primitive understanding of the concept. Several studies have demonstrated students' commitment to the notion of function only when represented by a formula (Markovits, Eylon, & Bruckheimer, 1986; Vinner, 1983.) The prevalence of "static", or pointwise, views of function has been noted by Monk (1987) and others. The difficulties students have in coordinating their function concept in algebraic and graphical representations has also been noted (Even, Lappan, & Fitzgerald, 1988; Resnick & Omanson, 1987; Schoenfeld, 1986).

Attempts to understand how students think about the limit concept suggest that informal translation by students of precise mathematical definitions creates difficulty (Tall & Schwarzenberger, 1978). In the area of continuity, Vinner (1987) found that despite being able to produce correct judgments about the continuity of functions, students' often gave answers based on incorrect reasoning.

Orton (1983a) has noted that students' routine performance on differentiation items was adequate, but that little intuitive or conceptual understanding of derivative was present. Amit and Vinner (1990) also have investigated the understanding of derivative, and found that students may equate the derivative of a function with the equation for the line tangent to the function at a given point. Although little research on the understanding of integration exists, Orton (1983b) found that technical facility could be quite strong, despite minimal conceptual understanding.

DESCRIPTION OF THE STUDY

PURPOSE

The purpose of the study was to develop descriptions of calculus students' understandings of function, limit, continuity, derivative, and integral, and to explore the interrelationships among their understandings of these conceptual areas. We were interested in the understandings developed by college calculus students as they experienced what might be described as a "traditional" calculus course, taught in a large lecture format, without use of technology.

METHODOLOGY

Over a two–semester period, six first–semester calculus students were selected as participants in the study. Although one of the researchers was teaching a section of calculus, the students were chosen from other lecturers' sections. We chose three students of each gender, and selected two with pretest scores that were below average, two with average pretest scores, and two with above average pretest scores. Each student was asked to participate in four interviews of about one hour's duration. The first interview for each student was audiotaped, and the remaining three interviews were audio-and videotaped. The interview topics were: function, limits and continuity, the derivative, and the integral. These interviews were spaced throughout the course, following the treatment of the topic within the course. The interviews were completed for four of the students. The other two students dropped the course.

INTERVIEW TASKS

The interviews were designed around a series of tasks developed by the researchers and/or adapted from the research of others. The intention was to provide tasks that help reveal students' ways of thinking about the central concepts of function, limit, derivative, and integral. Students were asked to complete the tasks and to "think aloud" as they did so. The interviewers interacted with probes and questions intended to help the interviewer understand the student's thought process.

The tasks chosen for this study include a mixture of items that are presented graphically and items that are presented via formulas. In each interview we included tasks that would be likely to be "routine" for the students, in that they were similar to homework problems in the text. A complete set of the interview tasks is available from the authors.

ANALYSIS

All of the interviews were transcribed. We began the analysis by studying the tapes and transcripts and seeking explanations for what occurred. Hypotheses and constructs available from the research of others helped to guide this process, and were used as general categories for sorting events. These categories were refined and reorganized through repeated study of the tapes and transcripts.

The interview series for one student, Sandy, has been analyzed in detail and is reported here. When appropriate, supporting examples from other subjects are also presented. Sandy's richest understandings and constructions were in the areas of function, limit, and continuity. Her conceptions of derivative and integral were very thin and defined mainly by procedural emphases.

RESULTS

The following descriptions are a preliminary attempt to characterize Sandy's understanding in the central areas of the study.

FUNCTION

Of all of the areas we studied, Sandy's understanding of function was the most thoroughly developed. A number of interesting features emerged . Three characteristics seem to define Sandy's image of function: algebraic formula, familiarity, and continuity. These characteristics embody certain contradictions, inconsistencies, and tensions. They also compete for priority in a context–dependent fashion. Note that the matter of values of the independent variable being paired with unique values of the dependent variable in no way figures into Sandy's thinking about function. Note also that matters of domain and range are not at all salient. The pointwise preference mentioned earlier also was apparent.

CHARACTERISTICS DEFINING FUNCTIONS

It is clear that use of the word "function" triggers, for Sandy, a search for an algebraic formula which defines the function, and in most contexts, a strong impulse to substitute certain numbers into this formula. When asked to provide examples of functions, she writes down equations for polynomial functions. When asked whether certain graphs in the Cartesian plane represent functions, a frequent criterion for deciding seems to be her judgment of whether or not there is an equation which is represented by the graph. For example, given the diagram in Figure 1 and asked whether this represents a function, she replies that

> I guess I'm not positive whether one function can be graphed and it would come out
> like that ... if the set of points that come out of this [graph] could be plugged into one
> function ... a formula that would make this ....

Figure 1

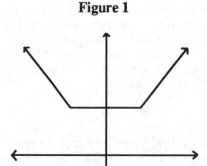

Later, when shown the graph in Figure 2, Sandy observes,

> It just doesn't look like it could be the graph of one equation.

She goes on to state

> I think of a function as having a rule that somehow coordinates or connects a set of ordered pairs.

Figure 2

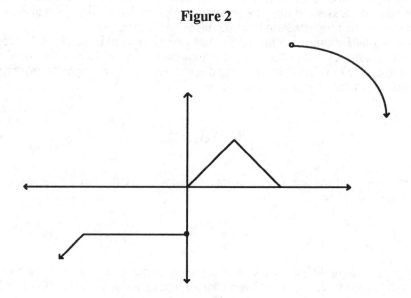

In response to the following question, Sandy reveals more about her notion of function. (She also reveals something about her conception of mathematics.)

TRUE OR FALSE: EVERY FUNCTION CAN BE EXPRESSED AS AN EQUATION.

> I think the answer is TRUE. But, maybe it's not because of the word every ... I think that there's some, like trig functions or something like that, that can't be expressed as an equation.

This same commitment to an "arithmetic" function is corroborated later when, after the interviewer asks whether she can give any other examples of functions, Sandy replies:

> ... Like a division one, or using a quotient, or like multiplying two numbers?

The need for an algebraic representation as an anchor and starting point is exceptionally strong for Sandy, and emerges repeatedly as she works on tasks in the other concept areas. This certainly is not a new finding and has been observed by others (Dreyfus & Eisenburg, 1983; Markovits et al., 1986.)

A competing criterion used frequently by Sandy in making judgments is whether the graph presented to her in a graphic setting is "familiar." Familiarity means having graphed curves like this before, or having seen similar graphs in class. When Sandy classifies a graph as "familiar," she automatically classifies this graph as a representation of a function. In Figure 3 are cases where familiarity led Sandy to classify the graphs as functions.

Figure 3

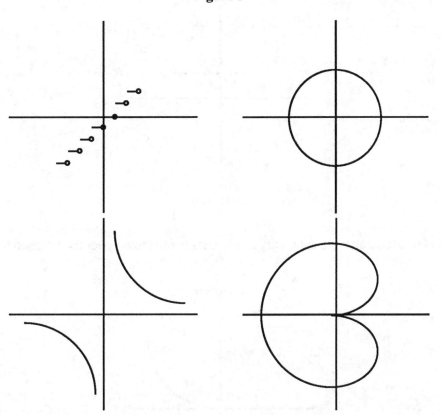

Sandy's third criterion for judging whether she is dealing with a representation of a function seems to be related to her concept of continuity, although she is inconsistent here. She will not agree that the graph in Figure 4 is a function,

> 'Cause it's just four separate points … it doesn't seem like they connect together in any way.

Figure 4

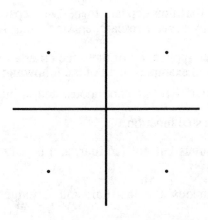

When asked what changes in the graph would result in a function, she says she would like to "connect the dots."

The example in Figure 5 prompted her to say

> I don't think it's a function … because of the separate dots … I don't think of that as
> a function.

Figure 5

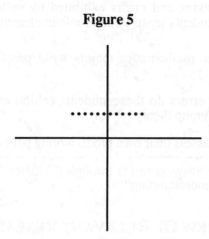

This apparent commitment to continuity also causes her to puzzle over the function represented by the graph in Figure 6.

Figure 6

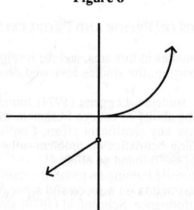

She is bothered by the fact that this appears to be two separate functions. (This difficulty with piecewise functions has been noted by several researchers, including Markovits et al. (1986), and Vinner (1987.)

We searched for a logic in her application of the three criteria of algebraic formula, familiarity, and continuity, but none emerged. We found examples of all of the following:

> A graph which is "familiar" but not continuous, and is judged (correctly) to be a
> function.
> (This task displayed a step function.)
>
> A graph which is continuous but not familiar, and is judged (incorrectly) to be a
> function.
>
> A graph which is continuous and familiar, and is judged (incorrectly) to be a
> function.

This suggests that all three concerns (algebraic formula, familiarity, and continuity) are important to consider in any curricular or instructional assumptions we make about what will make sense to students.

One might ask, rightfully, how important it is that Sandy be able to determine whether a graph or an equation represents a function. It is certainly possible to do a great deal of calculus successfully without this particular facility. The issue here may have to do with Sandy's receptiveness as a learner to explanations and problems that are presented in graphical form; if she does not accept as functions graphs that are not familiar, or are not continuous, or do not seem to have a formula, then it is not at all clear how readily she will engage with explanations and problems that hinge on interpretation of the graph. Further, it is unlikely that graphical representations will be useful tools for her in enriching her concepts of central calculus ideas which happen to have appealing geometric interpretations, such as derivative, limit, and definite integral. Finally, given current optimism about graphing packages and graphics calculators as promising tools for enriching calculus learning, it seems important to explicate further the ways in which this lack of coordination and expectation of an algebraic formula may impede learning, or make some of the more interesting calculator and computer–based tasks very alien for students.

POINTWISE PREFERENCE

Several researchers have articulated distinctions between pointwise and global means of interacting with functions. Monk (1987), in studying students' conceptions of derivative, talks about "across–time" understanding and "pointwise" understanding, and argues that beginning calculus students appear to operate most successfully from a pointwise view of function, whereby they consider the function by thinking about only one point at a time. This tendency did emerge to some extent with Sandy, and more strongly with Linda. For Sandy, her predilection for a pointwise rather than across–time interpretation seems to vary depending on the context. When working with graphical representations, Sandy seems relatively facile at using a global (across time, qualitative) interpretation, although in the algebraic setting this is difficult for her; she operates almost entirely in pointwise fashion.

Sandy, when asked to

DRAW A FUNCTION f SUCH THAT $f(1) = 2$ AND $f(3) = 6$

asked:

So it's just going to be one thing that I'm drawing, or two separate things?

She is unable to complete the task.
Both Sandy and Linda were completely stymied by the following task:

DRAW A FUNCTION f SUCH THAT $f(2) > f(-3)$.

Sandy gets no further than to notice that "the value of 2 is greater than the value of −3". Linda interprets that $f(2)$ and $f(-3)$ are both functions, and doesn't make any additional sense. It is possible that, because of the introduction of this task through an algebraic representation, the students are drawn to a "pointwise" method of operating, and thus cannot move to the more qualitative level necessary to complete these tasks. It also appeared that there may have been some difficulties in reading the notation.

THE REAL LINE

Although no tasks were developed to explore students' understandings of the real numbers, in several places in the interviews, evidence about their thinking emerged. Most striking was their lack of understanding of the properties of the real numbers. Sandy, in the limit interview, spoke of "being all the way to 0 except not right there" (an interesting contradiction.) Kevin describes points "just to the left" and "just to the right" of 0. Additional research is needed to relate the ways in which students understand the real numbers to their concept of limit. This study makes only preliminary forays into this question.

LIMIT

The limit concept is particularly interesting for several reasons. Much of the language traditionally used in introducing the concept *(approaches, as near as we like, gets closer to)* is natural language imbued with its own meaning. The students' understandings of the concept seem deeply intertwined with this natural language. Sandy was struggling with the question of whether $.\overline{9}$ is equal to . She argues that

You get as close as you can which would be .9999 ... but it wouldn't be quite 1 ...
for the work that we do in limits, .99999 ... would be close enough to solve the
problems that we needed to solve.

It appears that the traditional language used to explain the limit notion to students may be helping
Sandy to shape a very fuzzy concept of what is meant by finding the limit of a function at a particular point.
Other researchers (Tall & Schwarzenberger, 1978) have examined this phenomenon more closely. Somehow
what Sandy perceives as the appropriateness of approximation as a central feature in the limit concept
transfers to her views of answers to limit problems. These too are in some sense "close, but not quite there."
She compares finding limits to sketching graphs, and argues that the answers "weren't that specific." She goes
on to say that she prefers derivative problems because

It's more definite and there's more of a specific and concrete answer. [With limits] it
seems like you can't quite define it or quite, you know, get where you're trying to go
to.

Sandy, as well as Kevin and Linda, displays a hierarchy of strategies in approaching limit problems.
Without fail, if "plugging in" seems to work, that is the strategy of choice. The criterion for whether this is an
appropriate strategy seems to be tied to whether, in Kevin's words, the x–value "fits neatly" in to the
function. This algorithm is performed with little thought as to the meaning carried in the limit notation. The
students see problems such as

$$\text{FIND } \lim_{x \to 2} x^3$$

as straightforward functional substitutions. Sandy, when asked in several ways if she could relate her correct
solution to the above problem to a graph that was provided, was unable to do so.

For Sandy, this strategy of solving limit problems by finding the function's value at the appropriate
point seems to hold up fairly well in situations presented graphically. When given the function in Figure 7 and
asked about the $\lim_{x \to 4} f(x)$, she responded correctly with 8.

Figure 7

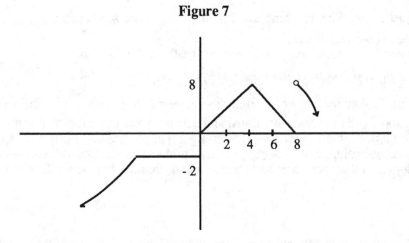

When asked about $\lim_{x \to 0} f(x)$, she answered "undefined" and gave some argument related to the
discontinuity at 0. Given her earlier equating of function with continuous, it may well be that her reason for
finding no limit is more related to not believing she has a function at 0. Unfortunately the interviewer did not
probe further here. However, Sandy was uncertain of her answer, and noted in some desperation that

Maybe if I knew what the function $f(x)$ was ... like a polynomial ... and you
plugged in the values

CONTINUITY

Sandy's concept of continuity is closely tied to the notion of a function's being defined. Sometimes,
her approaches to problems in the graphical setting and to problems in the algebraic setting, relative to the
concept, are quite similar mathematically. She basically seems to judge a graph to represent a continuous
function when there are no holes or jumps in the graph; a function presented algebraically is judged to be

continuous as long as it seems possible to substitute in any x without causing trouble. She does seem to equate a "jump" in the graph with a point at which the function is undefined.

In her understanding of continuity, Sandy's methods for determining whether a graph of a function is continuous are clear and consistent, and are described by her repeatedly in everyday English ("it just keeps going," "it all flows together," "it continues," "t keeps going this way.") Questions about the continuity of a graph at a single point caused her to question "whether a function can be continuous at a specific point." Not three minutes earlier in the same interview, she has explained that, when given a function defined by a formula, her method of determining "a point where that would make it not continuous" would be to determine where the "equation" is undefined. Thus she seems perfectly comfortable holding pointwise and global criteria for continuity at the same time, although these seem to generate some confusion for her.

Sandy's response to a request to sketch a discontinuous function is shown in Figure 8.

Figure 8

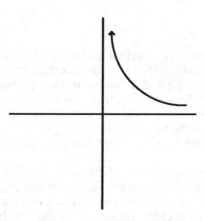

She argues that it is discontinuous because it is undefined at 0. She then admits, tentatively, that "maybe it's continuous from x equals 0 to infinity ... because it keeps going."

She then provides the example in Figure 9 and declares that this is "continuous up to a point", and then argues that the function is discontinuous because "there's a point where it's undefined."

Figure 9

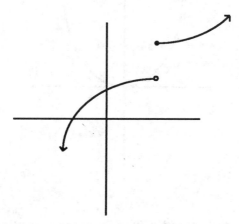

When asked about the continuity of the graph in Figure 10 at $x = 2$, she decides that the function is discontinuous there, but is perplexed by the fact that the function is defined at 2.

She actually reverts, at this point of confusion, to her need for familiarity.

> I've never seen a function like that graphed before ... I don't know how you would get a point up there in graphing a function.

She is not willing to engage with the problem because she is not convinced that she is dealing with a function. Perhaps her difficulty stems from not being able to imagine a formula that would yield such a graph.

Figure 10

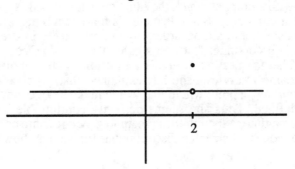

DERIVATIVE

Certainly in the area of derivative Sandy had the most difficulty in relating her graphical and formula–based understandings. Her facility in computing derivatives using algorithms was excellent. Her ability to sketch curves following an algorithm about testing points, looking at positive and negative derivatives, et cetera., was fine. Yet in the same interview she admitted to having no idea how the tangent line related to the derivative of a function. Nor would she venture any discussion about the way in which the derivative of a function was related to the function itself, despite being able to mouth the steps of a curve–sketching algorithm correctly. The connections between her procedural and conceptual knowledge are not at all strong here.

Neither global nor pointwise notions of "differentiability" have geometric meaning for Sandy. Certainly she displayed no evidence of any geometric interpretation of differentiability that might be related to the idea of smoothness. To make decisions about whether certain graphs represent differentiable functions, she reverts to her previous practice of seeking formulas for the functions, testing whether she can find a derivative by standard procedures, and, when she can, agreeing that the function is differentiable. In one task, the sketches in Figure 11 were provided, and she was asked to sketch the derivative of each function on the same axes.

Figure 11

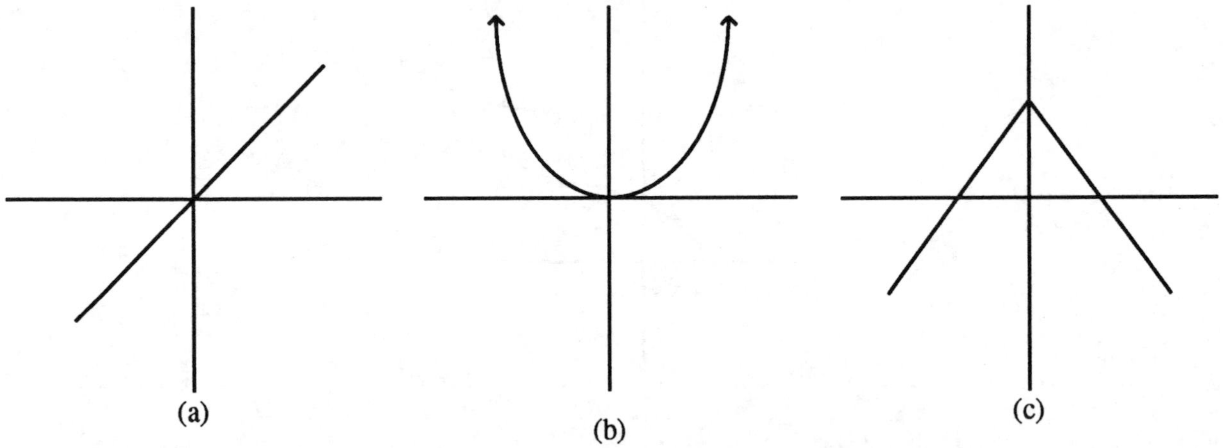

(a)   (b)   (c)

She accomplished this in parts a and b by determining equations for the functions, taking the derivatives, and sketching the derivatives. Her procedure in part c was more sophisticated —she was uncomfortable because the graph was "unfamiliar",but she actually determined reasonable equations for the two rays involved and then noted the slopes correctly.

Sandy was asked to calculate the derivative of $f(x) = 3x + 1$. She did so correctly. She was then asked how this derivative relates to the original function, and indicated that she was not sure. The possibility of two functions having a relationship to one another may well be at odds with her view of functions' as an algebraic formula into which numbers must be substituted. It is also quite possible that her general preference for dealing in a pointwise fashion with functions is interfering with her openness to thinking about the ways in which two functions might be more generally related. When asked directly whether the derivative of a function is a function, she responds

Yes ... because you can take the second derivative.

INTEGRATION

The integration interview was intended to check Sandy's interpretation of antiderivative tasks but then to move on to her understanding of the concept of the definite integral and the Fundamental Theorem of Calculus. The interview took an interesting turn when it became clear at the outset that she saw only inconsequential differences in the problems $\int x^2\,dx$ and $\int_0^2 x^2\,dx$. She said "They're both taking the antiderivative — but the second is more definite." As before, her procedural skills were strong — she readily found answers to basic problems. As usual, she demonstrated great reluctance to use geometric interpretations as a help in completing an algebraic process, and was much more inclined to move to her algebraic context when possible.

Sandy, as well as other subjects, interprets the integral as a signal to "do something." This is consistent with the view of function as a formula into which a value must be substituted, rather than an entity whose characteristics can be studied. Dubinsky and Schwingendorff (1990) describe this phenomenon as a "process, i.e., a procedure which transforms an input into some output" (p.176).

The most interesting feature of this interview was Sandy's explanation of the relationship between the limits of integration in a definite integral and the constant of integration in an antiderivative. This explanation seems to be related to her fundamental understanding that a definite integral "defines the area between the graph of the function and the x–axis." She later says, "when you do the antiderivative of a function, it's the area between the graph of that function and the x–axis."

Sandy has successfully solved the problem $\int_0^2 x^2\,dx$. She then is asked how the problem would change if it did not have any limits of integration. Here is what followed:

Sandy: You'd have a constant on the end because you wouldn't know that $x = 2$.

Interviewer: Find $\int x^2\,dx$.

Sandy: (writes $\int x^2\,dx = \dfrac{x^3}{3} + C$).

Interviewer: What does the C mean?

Sandy: This is like finding the area under the curve, but it doesn't define how far under that x–axis you have to go. So you put a C in. It could be any C or any constant. And also, there's another reason, too. 'Cause if you take the derivative of just a number, like 1 or 2, it's 0. So any number could be in there.

Interviewer: Can you draw a picture?

Sandy: I think this would just be the graph of x [Figure 12]. And, then the C is just … well, okay, it would be, say, between … the C could be anywhere along here. So that would be how far. Let's say $C = 2$, and 2 is here, so that would be how far to draw it. And the graph of the function and the area would be defined between that area down here.

Figure 12

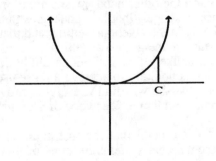

It is also likely that Sandy's notion of Riemann sums and the definition of the definite integral is at best fuzzy and not necessary to her routine solution of calculus problems. When shown the diagram in Figure 13 and asked whether this has anything to do with integration, Sandy answers:

> Yes, I think it's the proof of why integration works because it has something to do with the area of each of these little boxes, then you add them up and get the area under the curve.

Figure 13

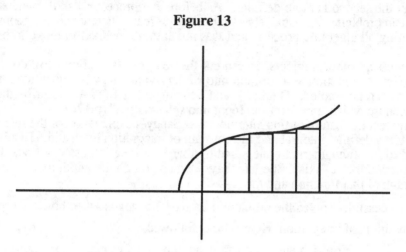

She is asked whether this is exactly the area under the curve, and replies:

> Not all the little boxes of area because you have this little space here where it doesn't touch, but for some reason I think there was something in the proof that makes up for that. But I'm not sure what.

This all culminates with the interviewer's question: "Do limits have anything to do with integration?" She responds:

> I don't really remember specifically, but it seems like there was something in the proof where you took the limit of the function … and you'd come up with an area and you'd add all of the areas together. I'm not exactly sure.

Clearly, despite great ingenuity in developing her own interpretations and building what, for her, are quite logical connections, whatever the instructor might have done in class to motivate the notion of the definite integral seems to have been lost.

GENERAL OBSERVATIONS AND IMPLICATIONS FOR PRACTICE

A number of observations emerge here that may have implications for additional research and for practice, particularly in the areas of instruction and of curriculum development.

1.   Graphical contexts and algebraic contexts may function for students as separate worlds, where unrelated algorithms for solving problems are applied, and where competing meanings can be held. Ability to coordinate the algebraic and graphical contexts may differ substantially across concepts.

In analyzing Sandy's interviews, evidence for this observation arose repeatedly. She made judgments about functions according to criteria that were mathematically different depending on the way in which the function was represented. Continuity, on the other hand, meant more or less the same thing for her in both contexts. There is even some evidence to suggest that her tendencies to think about functions in a pointwise fashion manifest themselves more vividly in the algebraic setting, and that tendencies to think about functions in an across–time fashion are more visible in the graphical setting.

In the case of Sandy, it seems unlikely that without carefully designed mathematical tasks that are sensitive to these ways of thinking, she is unlikely to achieve a more useful, coordinated view of these two rather separate worlds. Nor is it likely that she will find it useful to move to a graphical representation as a convenient strategy for solving certain problems. The work of Dubinsky and Schwingendorf (1990) and others offers promise in this area.

Cobb et al (1991) caution, "it is crucial that the instructional developer distinguish between the meanings that students give to representational systems in terms of their current ways of knowing and the mathematical structure that the system embodies for adults who know mathematics." (p. 5). This is advice that would be well–heeded in this area.

2.   Competing, conflicting conceptions and conclusions are held quite comfortably and routinely in the development of calculus concepts. In many cases these conflicts seem to exist between the formal, mathematically conventional meanings held by instructors and textbooks, and the personal and idiosyncratic sense–making that occurs with students.

There are several indications that Sandy and the other students were not bothered by contradictions between the formal definitions and their working definitions. Vinner (1983) has discussed this issue of "concept image." The use of personal, working definitions and "criteria for judging" that are different from what she was exposed to in class, and that are adequate in many (but not all) cases, was routine for Sandy. These definitions seem to be built around examples. She was very comfortable functioning without formal definitions in most cases, although occasionally realized that a definition might be helpful.

"Natural" language is used readily in describing mathematical ideas, and appears to be more useful than formal mathematical definitions or vocabulary. There are numerous examples through Sandy's transcript, as well as in the other interviews, where this is supported. Sandy's concept of continuity in the graphical setting is described repeatedly with language such as "it just keeps going", "has no definite end", and "it all flows together."

The level of Sandy's reliance on outside authority is uneven and interesting in that it relates to the depth of her engagement with mathematical tasks. It seems that the standard direct instruction that she experienced did not invite her to question or think deeply about very many things. For example, in the functions interview, in the course of discussing what makes something a function, she says:

> I don't know … we've been looking at functions in class … he writes a function of
>
> x, $f(x)$ equals … you know, whatever, and he tells us that's a function. So I just know that's a function, but I don't … I'm not really sure of … a definition of it or whatever.

3.   Calculus students will actively formulate their own theories, build their own connections, and readily construct meaning for problem situations. These processes seem to be influenced strongly by previous experience and knowledge. There are powerful tendencies to call upon familiar examples and frequently–used patterns.

The entire research base in precalculus and calculus learning makes this point most convincingly. With the benefit of varied methodologies, contexts, and research tasks, this finding emerges most clearly. The present study simply provides additional examples of this phenomenon, and these are discussed through the analysis section. Sandy was constantly forming her own connections, many of which were unexpected by the interviewer, and most of which were rational, based in experience, and adequate for certain problems.

With Sandy, it is fair to say that her approach to almost every task was first to try and recollect an algorithm or procedure for dealing with the task. There were few examples of where her initial strategy was to reason from any conceptual base of understanding.

The following example is an interesting case of Sandy bringing her search for algorithms, her mathematical experience in other settings, and her willingness to formulate her own connections all to bear in a mathematical task.

Sandy, when confronted with the exercise below, reasons that this must be true because "they show a positive and a negative number."

THE FUNCTIONS f AND g EACH CONTAIN INFINITELY MANY POINTS. HERE ARE PARTIAL TABLES FOR EACH:

x	$f(x)$		x	$g(x)$
0	0		0	0
−1	1		−1	1
1	−1		1	−1

TRUE OR FALSE: f AND g ARE EQUAL FUNCTIONS.

The practice of testing a point in each of several intervals as a step in determining where a function increases and decreases in graphing is part of the repertoire offered in many texts, and it is quite possible Sandy is generalizing from this setting to a rather different one. (Sometimes knowing about what happens at two different places is adequate for generalizing, and sometimes it isn't!)

4.   Startling inconsistencies exist between performance, particularly on procedural items, and conceptual understanding. Traditional means of assessment in calculus are almost certain to mask the nature of student understanding.

This study, as well as a host of others, has found solid evidence that students can perform the procedural tasks of calculus with rather astonishing success while displaying conceptual understandings that are not what we would like to have in place. The most striking example in Sandy's interviews was in the derivative interview. She was asked what the derivative means. In response she gave a lengthy and virtually flawless description of an algorithm for curve sketching, including the ways in which the first derivative of a function may be used to gain information about the function's graph. Not one minute later, in the same interview, when asked what tangent lines to curves have to do with derivatives, she replies

> I'm not exactly sure ... I can't remember exactly how it's related to the derivative ...
> I remember doing it, but I can't remember exactly how.

It seems that the availability of technology is a most promising factor in building curricula that could change this state of affairs. As the procedural tasks of calculus become less prominent in instruction, it would seem that whatever develops to take the place of procedural emphasis should attend to the development of solid conceptual understanding of central ideas, understanding that allows students to solve problems in new domains.

5.   Students build a conception of mathematics as a result of experiencing instruction.

Although this study was not designed to explore students' conceptions of mathematics, several pieces of interesting evidence in this area arose through the interviews and should be noted. Sandy displays an incompatible combination of faith and mistrust in the mathematics. She feels that

> You just have to practice the problems and then you understand,

yet observes that

> There's always, like, one exception or two exceptions to ... every rule that's
> supposed to happen all the time.

While struggling with whether or not certain functions are continuous, the interviewer asks whether she thinks there's a way to know definitely. She answers that she is sure there is, although it is clear from the context that she does not have this secret, and is cheerfully willing to forge on anyway.

She goes on to claim in a later interview that she feels secure about an answer when it agrees with the answer in the book, or when the problem is so similar to an earlier problem that she can rely on following the pattern.

In many instances, Sandy demonstrated conflicting conceptions held simultaneously — for example, that mathematics is reassuring because there always are concrete answers; yet in solving problems involving limits, she is not at all sure that the answers are "concrete."

Confrey (1990), in discussing the implications of constructivism for classroom instruction, contends that ultimately the students must decide on the adequacy of their own constructions The reliance on the outside authority of the instructor or the text must be supplanted by a validity that is internal to the students. For an instructor to facilitate this autonomy, and at the same time to guide students toward powerful constructions that agree with those of experts requires,that teachers build models of students' understanding, and enrich what Lampert (1988) describes as their "map of the territory."

CONCLUSION

The preliminary findings concerning the nature of Sandy's understandings in graphical and analytic contexts, and the difference in her understandings and the more common "publicly shared" conceptions, are important. The unevenness in coordinating graphical and algebraic representations is especially critical for those concerned with calculus. Although Sandy is certainly only one student, there is substantial evidence elsewhere that students in secondary school and beyond have limited facility in making connections between representations in the Cartesian plane and representations in the formal algebraic system. It also is fair to say that most traditional presentations of calculus include substantial graphical examples and use of pictures presumably to convey the essence of concepts. Resnick and Omanson (1987) provide negative findings in this regard. It may well be more complex than the rather simple problem of not making connections. If students' concepts, methods of reasoning, and repertoire of heuristics are radically different in each of these representations, and if students are comfortable with the inconsistencies, contradictions, and competing meanings that emerge as a result, then the challenge of helping them reach a workable means of connecting these representations is very complex. An additional variable, of course, is the part to be played by technology.

University mathematics faculty have a responsibility to help students understand central mathematical ideas in a way that will be useful to the students that opens up for them the beauty of the mathematics, and that empowers them to learn more mathematics. As the research base in calculus learning grows, and as the methodologies and perspectives become more accessible, perhaps all of these dimensions of research can play a useful role in assisting with that responsibility.

Hart provides a model approach to Research in Undergraduate Mathematics Education in his carefully documented study and extensive use of prior work. His central operating construct is the notion of conceptual schema. He states that "proof processes, errors, outcomes, and metacognition are all inextricably linked to, and substantially explained by, the stability, power, and accessibility of the conceptual schema possessed by the student. Moreover, it is proposed that process use, metacognition, and misconceptions are in fact part of a student's conceptual schema."

In particular, Hart's study addresses two issues: representation and general versus domain-specific strategies. His major point in this regard, that general strategies and domain specific knowledge interact strongly, is one with significant and broad implications in mathematics education. As Hart notes, this finding has been developed in other studies, which only heightens its import here.

The amount of data, on one hand, is insufficient to make significant regularity apparent. Six problems and less than three dozen students, while producing a sizable volume of raw data, uncovers a class of particulars that are undoubtedly part of a larger picture. Its structures and regularities require yet more data from more students and more problem situations. Such situations might also include constructive activities, where students build operation tables for groups, for example, or create mappings from one to another. Investigation of some of these sorts of activities may yield insights into the critically important process of schema *construction*. One way to get insight into the structure of a complex system is to watch it being built. Hart's paper is primarily an attempt to learn about the system by watching it in operation, during the process of proving given statements. Yet another way to get insight might be to put the student in the position of generating conjectures and building problems.

As with most studies reported in this volume, it is a beginning, an early step into a needed research program that promises major insight into a difficult problem area in undergraduate mathematics education. Moreover, as with other studies reported here, it reflects student knowledge built from what we have every reason to believe are currently common experiences in abstract algebra courses.

A CONCEPTUAL ANALYSIS OF THE PROOF–WRITING PERFORMANCE OF EXPERT AND NOVICE STUDENTS IN ELEMENTARY GROUP THEORY

Eric W. Hart
Maharishi International University

INTRODUCTION

A significant issue in collegiate mathematics education is the teaching and learning of proof. Everyone concerned acknowledges the difficulty of proof, but there is little research–based understanding of why it is difficult. It is the intent of this paper to shed some light on this important issue. It will do so by describing a research study conducted by the author on proof in the area of group theory. This study analyzes the processes, errors, outcomes, and self–assessment of college students, at four operationally defined levels of conceptual understanding, as these students attempt to write six carefully chosen proofs in elementary group theory.

In the process of presenting this study the extant literature concerning proof at the college level will be reviewed. The relevance of the literature reviewed to the group theory study will be discussed throughout. The review of the literature is not intended to be complete, but it is reasonably comprehensive.

THE IMPORTANCE OF PROOF

There is no doubt that proof is an important part of mathematics and something that should be taught. This has been attested to by, among others: Fawcett (1938) in his classic longitudinal study on proof; Polya (1954 and 1957) in his well–known books on problem solving; the Commission of Mathematics appointed by the College Entrance Examination Board in its important report of 1959; Baylis (1983), who states in the very title of his article that proof is the essence of mathematics; Driscoll (1983), who perhaps overstates that, "it [proof] is the fundamental tool for extending the field of mathematics" (p. 155); and Hanna (1983, 1989, 1990), who maintains the importance of proof even though she correctly questions its role as the all–encompassing methodology of mathematics. The recent influential document from the National Council of Teachers of Mathematics, *Curriculum and Evaluation Standards for School Mathematics* (1989), recommends teaching proof in several of its standards for high-school mathematics. Further evidence of the awareness of the importance of teaching proof is the appearance of "how–to" textbooks on proof, such as Watson (1978) in England, Solow (1989) and Cupillari (1989) in the United States, and Franklin and Daoud (1988) in Australia.

THE DIFFICULTY OF PROOF

Despite its importance, proof has been difficult to teach and learn. The difficulty is certainly not unique to this generation. Poincaré was astonished at students' inability to understand proofs in 1914. He wrote, "There is nothing mysterious in the fact that every one is not capable of discovery. That every one should not be able to retain a demonstration he has once learnt is still comprehensible. But what does seem most surprising, when we consider it, is that any one should be unable to understand a mathematical argument at the very moment it is stated to him" (p. 146).

More recently, at the high-school level, Williams (1980) found that, "Only those students who were classified as high achievers by their teacher, less than 30% of the sample, exhibited any understanding of the meaning of proof in mathematics" (p. 165). Usiskin and Senk (as reported in Senk, 1985) found that at the end of a full-year course in geometry 50% of the students can do at best only trivial proofs.

Although research at the college level is scarce, the results are similar. For example, Frazier (1969) found that on a criterion-proof test only 7 of 55 students enrolled in a liberal-arts freshman mathematics course received a score that was above 50%; and Martin and Harel (1989) found that of 101 preservice elementary teachers, "38% and 52% accepted an incorrect deductive argument as being mathematically correct for familiar and unfamiliar statements, respectively" (p. 41). Such results are not unique to the United States. Baylis (1983), in an informal assessment of British college students' understanding of proof, states that, "It is becoming increasingly apparent that the majority of students between the 'O' level and the end of first year of a degree course have only a nebulous idea of what a proof involves … " (p. 409).

DEVELOPMENT OF THE ABILITY TO PROVE

In the present study the proof–writing performance of college mathematics majors is examined, with an emphasis on describing differences among students at different levels of conceptual understanding. This developmental aspect of the study is important since psychologists and mathematics educators have long been interested in the development of mathematical thinking (e.g., see Lesh and Landau, 1983; Ginsburg, 1983). In fact, Lester (1983) states that, "… far too little is known about which processes individuals use naturally, when they acquire these capabilities …, and which processes individuals can be taught to use at a given stage of their development" (p. 237).

The two most prominent theories related to the development of the ability to prove are Piaget's theory of cognitive development and the much more specific van Hiele theory on the development of the ability to write and understand proof in geometry. The Piagetian theory is not discussed with respect to this study since all subjects in this study are well beyond the age normally associated with the acquisition of formal operations reasoning. (It should be noted, however, that studies such as Arons (1977) show that adults beyond the typical age for developing formal operations reasoning may nevertheless lack such reasoning ability; and, in any case, Piagetian–inspired teaching approaches can be used at the college level.)

The van Hiele levels model, particularly as it can be extended (see, e.g., Hoffer, 1983), is at least a priori relevant to this study because of the levels of conceptual understanding involved. Just how the design and results of this study can be viewed in the light of so–called "levels models" will be considered later in the discussion section.

RELEVANT RESEARCH PARADIGMS

Besides the van Hiele model, two other research paradigms are particularly relevant to the design and results of this study — the expert/novice paradigm and what Lesh (1985) calls "conceptual analyses."

Concerning expert/novice–type research, there is some disagreement among researchers concerning just how to define experts and novices. According to Schoenfeld (1982), an expert problem-solver is characterized not by proficiency at solving problems due to content mastery, but rather by the ability to use whatever knowledge is at hand to solve nonstandard problems. On the other hand, Lesh (1985), talks about "content–domain experts." These are people who have "acquired stable and accessible conceptual models for interpreting and manipulating information in that problem domain" (p. 325). Similarly, Owen and Sweller (1989) state that, "one of the determinants of expertise in a given domain is the possession of a large body of domain specific schemata" (p. 323).

The study reported here is concerned with how experts develop in the content domain of elementary group theory. That is, it considers "content–domain experts" and "content–domain novices."

A further refinement in the definition of "expert" is made by Silver (1985) who claims that there are not just experts and novices, but gradations between and within these categories. He maintains that we should "not act as if 'novice' and 'expert' were terms denoting equivalence classes of interchangeable elements" (p. 251), and that, in fact, we should "study individuals on their path to expertise, as they shed their misconceptions and naive beliefs" (p. 256). The present study takes a similar viewpoint, looking at the proof processes and misconceptions of college students who are at four operationally defined levels of expertise in the content domain of group theory.

The research paradigm most relevant here is the conceptual analysis model. The key points characterizing this model, as described by Lesh (1985), are: (1) "Interdependencies between content–understanding and process–use begin to emerge as salient" (p. 310); (2) "successful problem solvers tend to use powerful content–dependent processes rather than general (and weaker) content independent processes" (p. 324); and (3) "The processes that we have found to be the most beneficial to our students not only are the powerful and content–dependent variety, they contribute to both the meaningfulness and the usability of specific mathematical concepts. They are integral parts of the underlying conceptual models" (p. 324).

RESEARCH QUESTIONS

The present study modifies and integrates the conceptual analysis, expert/novice, and "levels" models in its attempt to describe the processes and errors exhibited by college students, at varying levels of conceptual understanding, as those students try to write proofs in elementary group theory. In particular, the following questions are investigated:

(1) How well do college mathematics majors write proofs in elementary group theory?

(2) What processes and errors do these students exhibit as they attempt to write proofs in elementary group theory?

(3) How do the students assess their own proof–writing performance?

(4) In what ways do the answers to (1) through (3) differ for students at different levels of conceptual understanding?

REVIEW OF RELEVANT RESEARCH

Three areas of relevant research will be briefly reviewed: (1) The effect of teaching heuristics on proof–writing and problem–solving performance; (2) The effect of other instructional methods on proof–writing performance; and (3) Process and error studies concerning proof. All studies reviewed are at the college level.

EFFECT OF TEACHING HEURISTICS ON PROOF AND PROBLEM SOLVING PERFORMANCE

There are very few empirical studies in this area, and the results of the research are mixed. While no claim is made here of a complete reporting, the studies reviewed do comprise at least most of the extant literature.

Concerning beginning college students, Leggette (1974) found that instruction in problem–solving heuristics improved the problem–solving ability of college freshmen, while Meyer (1982), in a similar study, found that such instruction did not have any significant effect. Conflicting results were also found in two studies concerning the effect of teaching heuristics on problem–solving ability in calculus. Lucas (1974) found some positive effect, while Ross (1980) found no effect.

With respect to the effect of heuristic training on proof, the results are again mixed. Cope and Murphy (1980) found that providing explicit instruction on a successful strategy for proving trigonometric identities resulted in improved proof–writing performance. Schoenfeld (1979) also found a positive effect for heuristic training, in this case on the proof–writing performance of upper-division college students. Goldberg (1974), however, investigated the effects of instruction in heuristics on the ability of non–mathematics majors in college to construct proofs in number theory, and found no significant differences among the treatment groups.

All of the studies reporting positive results have certain limitations. The Cope and Murphy (1980) study looked at a specific strategy for a specific type of proof (trigonometric identities) and thus has limited generalizability. In Cope and Murphy (1980) and Lucas (1974) the crucial issue of transfer is not addressed. In Schoenfeld (1979) transfer is positively affected by the treatment, but the study is limited with respect to sample size (seven) and generalizability from a laboratory setting to a classroom setting independent of the investigator.

It should be noted that the studies showing no effect for heuristic training also have limitations. As stated by Schoenfeld (1985), "In part, the reason for such discouraging results was that the complexity of heuristic strategies had been consistently underestimated; with hindsight it appears that adequate groundwork was not established for heuristic mastery" (p. 192). Further, these studies were all rather standard treatment/control studies, and it is now thought by many that the best way to study heuristics and to see their effect on students' learning and problem–solving ability is to take a more qualitative, cognitive science approach (see, e.g., Schoenfeld, 1987).

In any case, the positive effect of teaching heuristics on problem–solving ability is far from established. In fact, there is lively debate on the issue. On the one hand, for example, Lawson (1990) claims that, "there is encouraging evidence that training in the use of different types of general problem–solving strategies has positive impact on performance in both mathematics and other curriculum areas" (p. 406). On the other hand, Owen and Sweller (1989) state, "evidence that the teaching of heuristics is effective is very sparse. Most available evidence suggests that superior problem–solving skill does not derive from superior heuristics but rather from domain specific skill" (p. 327). Sweller (1990) goes on to say, "on the literature cited, we might be entitled to conclude that the evidence [that the teaching of heuristics is effective], rather than being sparse, is effectively non–existent" (p. 413). The present study is designed to shed some light on this important issue.

EFFECT OF OTHER INSTRUCTIONAL METHODS ON PROOF–WRITING PERFORMANCE

Several researchers have looked at the effects of methods other than heuristic training on proof–writing performance. The most common method investigated is teaching formal logic, although it is interesting to note that all studies in this category seem to be from only a brief period of time during the New Math movement. A few studies on the effects of logic showed positive results, like Howell and Melander (1967) and McCoy (1971), but most showed no effect, like Frazier (1969), Macey (1970), Williams (1971), and Walter (1972).

One study in this category, Frazier (1969), had to do explicitly with elementary group theory. The experimental group that received explicit instruction in formal logic and how to apply it to writing proofs in elementary group theory did significantly better on a test which gave students a small axiomatic system and asked them to prove three elementary theorems. However, on all three of the other parts of the criterion post–test — which included tests for knowledge of groups, knowledge of logic, and ability to write proofs in elementary group theory — there were no significant differences.

An anecdotal comment further reinforcing the research evidence that teaching logic has little effect on improving proof–writing performance comes from an error-analysis study reported below. In that study Selden and Selden (1987) state that, "we note that there is remarkably little correlation between the reasoning errors we have observed and the topics emphasized in an introductory logic course or even in one of the newer courses as on transitions to advanced mathematics" (p. 469).

Other instructional methods investigated with respect to proof include abstract versus concrete approaches (Caruso, 1966), guided discovery (Patterson, 1969; Espigh, 1974), and remediation (Burger, 1972). Little or no positive effect was found in these studies.

In more qualitative studies, Dubinsky (1986, 1988) devised and analyzed a Piagetian approach to teaching mathematical induction whereby the learner uses reflective abstraction to construct new schemata out of old ones; and Leron (1991) investigated the use of the programming language ISETL to teach abstract algebra, including proof. The results from both of these studies were encouraging. Concerning the issue of transfer discussed in the last section, the latter study by Dubinsky reports that a certain amount of transfer did take place.

Of course, outside of research studies, many instructors have tried many methods to teach proof. A brief sampling of methods reported in the literature follows.

Many instructors advocate some version of the so–called Moore Method (e.g., Cohen, 1982), and there has been some research into Moore's original method. Another common method is to offer a proof seminar concurrent with other mathematics courses (e.g., Reisel, 1982; Marty, 1989; Keny, 1990). Leron (1983) recommends structuring mathematical proofs in a manner similar to the "top–down" programming methodology of computer science. Leron (1991) and Dubinsky and Leron (1991) have worked on teaching abstract algebra by having students write programs in ISETL. Baylis (1983) claims that a teacher can make the learning of proof easier by stressing the wide–ranging power of a few simple principles, like "choosing the least element." A method whereby students are given alleged proofs which they are to grade for correctness and clarity is described by St. Andre and Smith (1978). As a final example, Movshovitz–Hadar (1988) outlines more "stimulating" ways to present theorems as a means of better setting the stage for the proofs. It is encouraging that the anecdotal evidence presented by the advocates of these methods indicates their positive effect on students' proof–writing performance. However, there are still serious questions that need to be answered regarding the instructor–independence of the methods and their generalizability to many students.

The lack of conclusive research–based evidence for teaching logic or using other instructional methods to improve proof–writing performance indicates that a deeper examination of students' proof needs to be undertaken. The present study, as well as studies in the next section, attempt to do this.

PROCESS AND ERROR STUDIES CONCERNING PROOF

There is a notable paucity of studies examining processes and errors related to proof at the college level. The present study falls partly into this category. This is such an important area that four other studies found in this category will be briefly reviewed.

Schoenfeld (1979) studied seven upper division college students--three in a control group and four in an experimental group. The instruction for both groups consisted of five sessions on problem–solving spread out over two weeks, during which time the students worked on and saw the solutions to twenty problems (many of which were proofs). The experimental group received instruction on five specific strategies that would help them solve the problems, while no explicit mention of heuristics was made in the control group. Through an analysis of think–aloud interviews the following conclusions were reached: (1) if we expect students to use heuristics, then we must teach heuristics; (2) the fact that students can master a particular problem–solving technique is no guarantee that they will actually use it; (3) when problem–solving strategies are taught and students think to use them, then the impact on problem–solving performance is substantial.

In another study, Schoenfeld (1982) set out to document the results of an intensive one-month problem–solving course, and to describe the instruments devised to measure the results. The subjects were twenty freshman and sophomores. The instruments were intended to be paper/pencil alternatives to think–aloud procedures. The first instrument measured how many plausible approaches were considered and pursued, along with how much progress was made. The second instrument was a questionnaire designed to elicit students' own self-assessment of their problem–solving performance. The final instrument measured heuristic fluency and transfer. Schoenfeld concluded that (a) these three instruments are easily-graded, reliable measures of students' problem–solving processes, and (b) a month–long intensive problem–solving course can have a positive impact on students problem–solving performance.

Selden and Selden (1987) described errors and misconceptions exhibited by college students as they attempted to write proofs in a junior-level abstract algebra course. The abstract algebra courses used for the study were those taught by the investigators at several different universities. Seventeen errors were documented and analyzed. The errors were classified in two ways: (1) according to whether or not they were misconception-based, and (2) according to their logical characteristics. In discussing their results, the investigators raise some questions concerning such things as the completeness of their list, the source of the errors, and how they might be prevented. They also suggest that students' reasoning abilities could be improved if lower-division mathematics courses would include instruction on creating and validating algorithms, rather than just implementing algorithms.

The final study to be mentioned here was a type of error analysis done by Martin and Harel (1989). In this study preservice elementary-school teachers evaluated existing "proofs". It was found that many accepted inductive arguments as valid mathematical proofs, many were influenced by the appearance of the argument rather than by its correctness, and many seemed to possess inductive and deductive proof frames simultaneously.

SUMMARY OF RESEARCH REVIEW

This review of research leads to three broad conclusions. First of all, although many instructional methods have been tried, we do not yet have a consistent body of research pointing to effective methods for teaching proof. Proof continues to be a problem–solving task at which most students fail.

Secondly, there is a need for more qualitative, cognitive–based research. It is still the case, as Shaughnessy stated in 1985, that "researchers ... need to carefully analyze the reasons for people's failure at problem–solving tasks in order to anticipate and overcome those potential failures when they build problem–solving models or develop instructional materials" (p. 400). More generally, we need to know more about students' thinking, about how they make sense of mathematical concepts and tasks. Such analysis has barely begun in the area of mathematical proof at the college level.

The third broad conclusion that emerges upon examination of the research literature is the lack of a coherent and unifying theoretical context for studying proof. As pointed out by Kaput (1987), such a lack "may be of critical importance to the efficacy of empirical work and its ultimate impact on the curriculum and instruction" (p. 19). Some progress is being made, as evidenced, for example, by the new interest in constructivism, but, again, much work needs to be done.

The present study hopes to make a contribution to the literature by taking a holistic, unifying approach to the study of students' ability to write proofs. In this approach, processes, errors, and metacognition are seen to be interdependent and fundamentally linked to students' conceptual schemata. This viewpoint will be further explained as the method and results of the study are presented.

METHOD

SUBJECTS

The purpose of this study was to analyze the processes, errors, and self–assessment that college students, at varying levels of conceptual understanding, exhibit as they attempt to write proofs in elementary group theory. The subjects were twenty–nine college mathematics majors from three progressively more advanced abstract algebra courses at a large state university. The first course was an introductory undergraduate course, the second was an advanced undergraduate course, and the third was a first-year graduate course. Ten students came from each of the first two courses, and nine came from the graduate course. All were (modestly) paid volunteers.

INSTRUMENTS

Two test instruments were used. The main instrument was the Proof Test, which consisted of six carefully selected proofs from elementary group theory; the other was the Proof Questionnaire, a questionnaire designed to assess the subjects' own perceptions of each proof task.

The Six Proofs

The proofs were designed to (a) require a minimal amount of factual knowledge, (b) reflect understanding of elementary group theory, (c) be representative of basic types of proof in algebra, (d) be do–able in 15 minutes each by students with some understanding of the definitions of group, subgroup, and abelian group, and (e) be complex enough to generate interesting data concerning processes and errors from students with a wide range of conceptual understanding. To design such proofs was not an easy task. They evolved over the course of two pilot studies and many tryouts with students and professors.

The factual knowledge needed for the proofs was that which was covered in the first few weeks of an elementary course, and, moreover, all relevant facts were supplied to the students on a "Prerequisite Knowledge" page. So, although the students certainly varied with respect to conceptual understanding, they were "equal" in terms of access to necessary facts.

On the questionnaire students were asked to rate their familiarity with each proof. If a proof was rated as "I've seen this particular problem before, and I remembered how to do it," then it was not included in the data analysis. Only eight out of the total 174 proofs had to be omitted for this reason. Furthermore, the Proof Test was administered later in the semester, at a time when students were not directly studying the topics on the test. Thus, any direct familiarity that students may have had with the proofs used in the study was minimized.

The proofs were chosen to represent types of proofs that frequently occur in algebra, as follows:

1. *satisfy axioms proof* — Proofs 1 and 6, where one has to prove that something is a group. Proof 6 is identical to Proof 1, but it is couched in an abstract setting.

2. *set–definition proof* — Proof 2, where one has to prove that a particular subset, given by a defining property, is a subgroup.

3. *uniqueness proof* — Proof 3, where one has to prove the existence of a unique idempotent element.

4. *syntactic proof* — Proof 4, where one uses a syntactic, i.e., procedural, "symbol–pushing", process to prove that a given group is abelian.

5. *non–routine proof* — Proof 5, where one has to prove that a group with an even number of elements has an element that squares to the identity.

The six proofs were:

Proof 1: Prove that the set of all positive rational numbers ($\neq 0$) with the binary operation Δ, where Δ is defined by: $x\Delta y = (x \bullet y)/2$ [$\bullet$ is standard multiplication of rational numbers], is a group.

Proof 2: Prove that, if G is a group and a is an element of G, then $\{x \text{ in } G \mid xa = ax\}$ is a subgroup of G.

Proof 3: If S is a set and $*$ is a binary operation on S, an element in S is defined to be an *idempotent* element if and only if $x*x = x$. Prove that in a group there is a unique idempotent element.

Proof 4: Prove that, if G is a group (with identity element e) such that $x^2 = e$ for all x in G, then G is abelian.

Proof 5: Prove that, if a finite group G (with identity element e) has an even number of elements, then there exists an element $a \neq e$ in G such that $a^2 = e$.

Proof 6: Suppose that a given set G with a given binary operation $*$ on G forms an abelian group. Define another binary operation, $\lozenge$, on G as follows: $a \lozenge b = a * b * t^{-1}$, where t is a fixed element of G and t^{-1} is the inverse of t in G under $*$. Prove that G with the operation $\lozenge$ is a group.

The Proof Questions

The other test instrument used in the study was a brief questionnaire designed to measure the students' own assessment of each proof task. After each proof the student was to choose the best response (multiple choice) to the following five statements/questions, adapted from Schoenfeld (1982):

1. Please rate your previous knowledge of this problem.

2. Please rate the difficulty of this problem.

3. Did you have an idea of how to start the proof?

4. Did you plan before you started the proof, or did you just "plunge in"?

5. Was your work organized or disorganized?

PROCEDURE

The main study was preceded by two pilot studies — one exploratory study conducted during an abstract algebra course that the author taught, and one "dry run" of the main study using five students.

For the main study, students were tested by the investigator either in groups or individually. At each testing session the proofs were distributed and attempted one at a time, and each proof was followed by the Proof Questionnaire. Fifteen minutes were allocated for each student to work on each proof, then, if he or she was not finished, 2–3 minutes were allowed for the student to briefly write down what he or she would do to finish, and then 2–3 minutes were used to fill out the questionnaire. Then the next proof was administered, until all six had been completed. The subjects were specifically instructed to write down what they were thinking as they attempted each proof.

The Levels Model

In order to address the issues of expert/novice and development of the ability to prove, four levels of conceptual understanding were operationally defined. These levels then provided the principal context in which the data were analyzed.

Students were classified into one of four levels of conceptual understanding of elementary group theory based on the correctness of their solutions to Proofs 4, 1, and 6. For the purposes of this study, elementary group theory was taken to consist of the definition of a group along with immediate extensions, such as subgroups and abelian groups. Proofs 4, 1, and 6 were used to stratify the subjects because each of these proofs requires a different level of understanding of elementary group theory — as determined by pilot testing and the following a priori reasoning.

Proof 4 requires only a "syntactic" understanding in that only "symbol-pushing" is required. This proof may look more sophisticated than Proof 1 or 6, but to complete the proof one simply sets up an equation and proceeds to manipulate inverses and the identity until one arrives at $xy = yx$. Thus, only procedural understanding is required, rather than meaningful conceptual understanding. Proof 1, on the other hand, requires a meaningful or "semantic" understanding of the defining properties of a group. In particular, knowing what identities and inverses are, and not just how to manipulate them, is required. Proof 6 is isomorphic to Proof 1, but it is couched in an abstract setting. Thus the level of understanding required for Proof 6, which can be called "abstract semantic understanding," is higher than that required for Proof 1.

Thus, correctness of Proofs 4, 1, and 6, in that order, served as the criterion for the levels of conceptual understanding. In particular, the levels are as follows:

Level 0   *pre-understanding* : none of the three criterion proofs were correct;

Level 1   *syntactic understanding* : Proof 4 was correct, but Proofs 1 and 6 were incorrect;

Level 2   *concrete semantic understanding* : Proofs 4 and 1 were correct, but Proof 6 was incorrect;

Level 3   *abstract semantic understanding* : all three criterion proofs were correct.

All but three of the students fit into one of the levels. Those that did not fit were classified as "non-fitters" (NF). The exact distribution of the students among the four levels is shown in Table 1.

Table 1
Distribution of Subjects Among the Levels

Level	Beg. Undergrad.	Adv. Undergrad.	Grad.	Total
L0	8	0	0	8
L1	1	2	0	3
L2	0	3	1	4
L3	1	2	8	11
NF	0	3	0	3

It is worth noting that students in more advanced classes were generally in a higher level, but not always. In particular, one of the beginning undergraduate students was at the top level of understanding, one of the graduate students was not at the top level, and the students in the advanced undergraduate course were fairly evenly distributed among levels 1, 2, 3, and NF. It is clear that, for the students in these classes, the amount of academic experience with abstract algebra does not necessarily reflect the level of understanding.

Data Analysis

Data were gathered on (a) processes used, (b) errors committed, (c) correctness of proofs, and (d) the subjects' own assessment of the proof tasks. Self-assessment was analyzed based on responses to the Proof Questionnaire. Correctness was scored using a method adapted from Malone, et al (1980), whereby each proof was graded on a scale of 0 to 4 as follows: 0— noncommencement; 1— approach made; 2— substantial progress; 3— result achieved with only minor errors; 4— completion. Processes and errors were investigated by analyzing the students' written work. Each proof was coded using an initial list of anticipated observable outcomes which was then modified and developed into a precise coding dictionary. To check the reliability of the coding, eighteen proofs — three of each of the six proofs — were randomly selected to be coded by an independent coder. Interrater agreement was .85 for processes and errors and .89 for correctness.

All data were analyzed across the four operationally defined levels of understanding. This appears to lead to some difficulty since three of the proofs were used both as the means for stratifying the subjects into levels and as instruments for collecting data. In fact, there is some difficulty with respect to the analysis of correctness. Any conclusion that subjects at higher levels had higher correctness scores is not very profound, since they were placed in higher levels based on the correctness of three of the six proofs. However, correctness of proofs is not a key variable in this study. It is included for completeness, but the focus of the study is on processes, errors, and metacognition at different levels of conceptual understanding, as discussed below. In these areas of primary concern there is no conflict between the means for stratifying subjects into levels and the subsequent data analysis across those levels.

RESULTS

The results will be presented in each of the four areas — correctness, self–assessment, processes, and errors — across all four levels of conceptual understanding.

CORRECTNESS

The mean scores on the proofs are presented in Table 2. Recall that the proofs were scored on a scale from 0 to 4, with a score of 3 or 4 indicating a correct proof.

Table 2
Mean Scores on the Proofs*

	Level 0 (n=8)	Level 1 (n=3)	Level 2 (n=4)	Level 3 (n=11)	Non–Fitters (n=3)	Over all Lvls. (n=29)
Pf 1	1.12	1.34	4.0	3.73	3.0	2.69
Pf 2	.62	3.0	4.0	3.27	2.33	2.55
Pf 3	2.37	2.34	3.25	3.55	3.33	3.03
Pf 4	1.0	4.0	4.0	4.0	2.0	2.96
Pf 5	1.12	2.67	1.5	3.27	2.67	2.31
Pf 6	1.0	2.0	2.0	3.73	2.33	2.41
All Pfs	1.21	2.56	3.12	3.59	2.61	2.66

*The proofs were scored on a scale 0–4. The minimum score indicating a correct proof is 3.

The key points from Table 2 are: (1) Higher levels generally had higher mean scores (no surprise, as discussed above); (2) the order of difficulty for the proofs, from easiest to hardest, was 3, 4, 1, 2, 6, 5; (3) for only two of the six proofs was the mean score over all levels up to the minimum correct score of 3; and (4) the mean scores over all proofs at levels 0, 1, and Non–Fitters were below the minimum correct score of 3.

SELF–ASSESSMENT

Table 3 summarizes the data from the Proof Questionnaire, and also includes a column of mean outcome scores (identical to the last row of Table 1).

Table 3
Mean Scores on the Proof Test and the Proof
Questionnaire Over all Proofs at Each Level*

Level	Outcome† (0→4) more correct	Familiarity (3→5) more familiar	Difficulty†† (1→5) easier	Start (3→5) better start	Planning (3→5) more planning	Organize (3→5) more organized
L0	1.2	3.2	2.6	3.9	3.8	3.8
L1	2.6	3.5	3.7	4.3	3.6	4.1
L2	3.1	3.5	3.7	3.8	3.5	3.8
L3	3.6	3.5	4.3	4.4	3.7	4.3
NF	2.6	3.5	3.7	4.4	3.9	4.0

* Mean scores on the Proof Questionnaire were calculated by assigning point values to the letter responses as follows: A=5; B=4; C=3; D=2; E=1. So, for example, the Familiarity question had scores ranging from 3 to 5 since only A, B, and C were possible responses; while the Difficulty question had scores of 1 to 5 since A through E were possible responses.
† Outcome scores less than 3.0 reflect unsuccessful mean performance on the Proof Test at the given level.
†† A Difficulty score larger than 3.0 reflects an average self–assessment, over all subjects at the given level and over all proofs, that the Proof Test, although possibly difficult, was done correctly.

The most interesting result from Table 3 is that Level 1 and Non–Fitters have Outcome scores that indicate incorrect proofs and yet the Difficulty scores indicate a self–assessment of correct proofs. At levels 0 and 3 this inconsistency was clearly not present, and at level 2 it appears not to be present. Other key points from Table 3 are: (1) generally the scores increase in all categories as the levels get higher; (2) although the scores for planning indicate not much planning at any level, the planning score was higher at level 0 than at the other levels; and (3) most subjects rated the proofs as unfamiliar.

PROCESSES

Processes used by students were analyzed in detail for each proof. The processes were subdivided into two categories: main and subsidiary. Main processes were those used for the entire proof or for each part of the proof if the proof had several distinct parts, as in the parts of showing something is a group. Subsidiary processes were those processes not classified as main. All results were carefully and thoroughly tabulated, but, since a full report would be prohibitively long, only a few of the most interesting results will be summarized here.

1.   In general, more main processes were used at level 0 than at any other level. However, some of the processes used were ineffective.

2.   It was rarely the case that some processes were absent at some levels. All processes seemed to be available at all levels.

3.   Different processes were emphasized at different levels. For example, "Guess and Check" and "Work Forward" tended to be used more often at the higher levels, while "Work Backward" was used more often at lower levels. "Work Backward" is generally found to be a more common strategy for novices (Owen and Sweller, 1989).

4.   "Reformulation" was more common at the higher levels. For example, for Proofs 4 and 5 it was necessary to consider elements, x, such that $x^2 = e$. It was much more common at the upper levels, rather than at the lower levels, for students to reformulate $x^2 = e$ as $x = x^{-1}$. This reformulation proved to be instrumental in correctly completing the proof.

5.   The process "introduce notation" was much more common at the higher levels than at the lower levels. It was used on only 23% of all proofs done by level 0 subjects, while 44%, 46%, and 48% of the proofs done by subjects at levels 1, 2, and 3 respectively, made use of the process. On Proof 6, "introduce notation" was used only once over all the proofs done by subjects at levels 0 and 1, while it was used on nearly half the proofs done by subjects at levels 2 and 3. As an example, recall that on Proof 6 one assumes that $(G,*)$ is an abelian group and it must be shown that $(G,\lozenge)$ is also a group where $a\lozenge b$ is defined to be $a*b*t^{-1}$, for a fixed t in G. Students at the higher levels were much more likely to introduce $e_\lozenge$ as notation for the identity under $\lozenge$. This proved to be a pivotal step for many correct solutions, and the lack of such notation led to mistakes in many incorrect solutions.

6.   There was often an inconsistent trend across levels. That is, the trend between levels 0 and 3 did not consistently evolve through levels 1 and 2. This could happen in many ways. For example, the percentage of students using the subsidiary process "change plan" over all six proofs was 25.0, 37.5, 54.2, and 16.7 for levels 0, 1, 2, and 3 respectively. Thus, between levels 0 and 3 the trend was for the frequency of use of "change plan" to go down, while for levels 1 and 2 the trend was up. (Note also, as another interesting result indicated by this data, that "change plan" was used least at level 3).

ERRORS

As with processes, errors were carefully analyzed and tabulated. Some of the most interesting results are reported here.

1.   The most common errors were "operation confusion" (Proofs 1 and 6), "incorrect deduction" (Proofs 2, 3, 4, and 5), and "assuming the result" (all proofs except Proof 3). Examples follow.

"Operation confusion" often occurred in proofs dealing with finding an identity element. In Proof 1, one had to show that (Q^+,Δ) is a group, where $x\Delta y$ is defined by $x\Delta y = xy/2$. A common error was to assert that 1 is the identity for this group, indicating a confusion between the operation Δ and standard multiplication in Q^+. In Proof 6, where $a\lozenge b = a*b*t^{-1}$, a common error in finding the identity for $\lozenge$ was to reason that if $a\lozenge e = a$, then $a*e*t^{-1} = a$, which implies $at^{-1} = a$, and so on (leading to an impasse or further errors). Here e, which in the statement $a \Diamond e = a$ is supposed to be the identity with respect to the operation $\Diamond$, is treated as the identity with respect to the operation $*$ when $a * e$ is collapsed to a.

"Incorrect deduction" was coded when, for example, abelian was deduced from the equation $ex = xe$. An instance of "assuming the result" was when students would "prove" that the identity exists by stating, "$a \Diamond e = e \Diamond a$, therefore the identity exists."

2. "Nonsense language," such as the statement "show $ax = xa$ is a group," occurred only at levels 0 and 1.

3. The percentage of proofs on which the process "correct error" was used was similar at all levels. Since there were fewer errors at the higher levels, this result indicates that students at the higher levels were more sensitive to detecting and correcting errors.

DISCUSSION

The discussion is mainly concerned with an analysis of the results with respect to conceptual schemata. The expert/novice paradigm, levels models, and teaching implications will also be briefly discussed.

CONCEPTUAL SCHEMATA

First a comment about terminology. The "conceptual schema" possessed by an individual is taken here to be similar to what is meant by that individual's "conceptual understanding", but the former phrase emphasizes an internal, individually constructed conceptual system, rather than an assimilated external, a priori conceptual system. In short, a conceptual schema is the evolving mental conceptual map that a learner constructs and uses to apprehend objects of knowledge.

The major premise to be argued here is that the conceptual schema possessed by a student is the key factor in determining success in problem–solving. That is, in the context of this study, proof processes, errors, outcomes, and metacognition are all inextricably linked to, and substantially explained by, the stability, power, and accessibility of the conceptual schema possessed by the student. Moreover, it is proposed that process use, metacognition, and misconceptions are in fact part of a student's conceptual schema. These premises are supported by the data of the study, as will now be discussed.

Processes and Errors

Some of the key results reported in the last section were that students at the higher levels tended to use the processes "reformulation" and "introduce notation" more often, while students at the lower levels committed the error "operation confusion" more often. All three of these results can be viewed in terms of students' conceptual schemata.

"Operation confusion," the most common error on Proofs 1 and 6, is clearly linked to an understanding of the generality and abstraction of the definition of a group. An overreliance on known concrete examples of groups, for example, can produce the operation confusion which results in claiming that the identity is 1, or that the inverse is the negation. This incomplete conceptual schema is described by Young (1982) as follows: "Students whose only knowledge of mathematical objects is through their concrete representations will be limited in their ability to assimilate new relationships which transcend the properties of those particular models" (p. 130). Another instance of the "operation confusion" error, again linked to an unstable conceptual schema, was the inability to correctly translate the group axioms in terms of the newly defined operation. This difficulty with translating defining axioms was also found by Espigh (1974) in the context of the axioms for a field.

THE PROBLEM OF REPRESENTATION

These errors— the inability to translate and an overreliance on concrete representation, can be viewed as errors of representation. The importance of meaningful problem representation in mathematical problem–solving has been found by many researchers (e.g., Wayne, 1981; Bull, 1982). Further examples of problem representation in the present study include use of the processes "reformulation" and "introduce notation," used predominately by higher level students.

"Reformulation" occurred in many guises. The reformulation of the group axioms in terms of the defined operation has already been mentioned. As another example, $a^2 = e$ was reformulated as $a = a^{-1}$ to provide the key to most successful proofs of Proof 5. While this reformulation could have come (syntactically) from simply multiplying both sides by a^{-1}, it seems more plausible from the context of the problem and students' work that it was driven by an understanding of inverses, in particular an understanding that any element multiplied by a which gives the identity must be a's inverse.

Another example of such a concept–driven strategy was the following instance of "introduce notation." In Proof 6, the notation $e_\Diamond$ was used to represent the identity for the operation $\Diamond$. The use of such notation seems plausibly driven by a stable understanding of the fact that group structure is determined by the particular operation. This link between notation and conceptual schemata is alluded to in Kaput (1991), where he states that mathematical notations are critically important as "tools to structure and support our own thought processes."

There is no doubt that representation is fundamental in mathematics, and thus the ability to represent is fundamental for doing and learning mathematics (Janvier, 1987; Kaput, 1987, Harel, 1991). The results of the present study show not only that students at lower levels exhibited errors related to representation, but also that students at higher levels exhibited successful representation–type processes. This distinction between higher and lower-level students with respect to problem-representation is well documented in the literature (e.g., see Schoenfeld, 1985, p. 244), but the point to be emphasized here is the link to conceptual schemata. It is the stability and accessibility of a student's conceptual schema that is often the driving force behind the crucial ability to represent problems in different, more useful ways.

GENERAL VERSUS DOMAIN-SPECIFIC STRATEGIES

Thus far representation has been discussed. We will now look at process use in general. The main question to be considered is: What are the differences in process use between higher-and lower-level students? One answer to this question is given by Sweller (1990), "The work on expert–novice differences led directly to the hypothesis that expertise consisted of a large store of domain–specific knowledge and strategies ... Little evidence was available of experts using sophisticated general problem–solving strategies not available to novices. At this juncture, it became reasonable to suggest that domain–specific rather than general strategies differentiated experts from novices." (pp. 411–12). The answer indicated by the results of the present study is similar, but a quite different viewpoint is suggested.

The critical issue is the distinction between general strategies and domain–specific strategies. Sweller emphasizes this distinction, as does Lesh (1985) in his explication of the conceptual analysis paradigm. How is the distinction defined? Different researchers have different ideas. Sweller gives an example of his view of a domain–specific strategy: "If you need to isolate the first pro–numeral of an equation of the form $(a+b)/c = d$, first multiply out the denominator" (p. 411). In contrast, Lawson (1989) states that, "Domain–specific strategies include heuristics, such as means–ends analysis ..." (p. 404). Schoenfeld (1985, p. 72) seems to delineate three levels of strategies: general heuristics such as means–ends analysis, subject–matter techniques, and, in between these two, the heuristics described by Polya, such as examining special cases.

Thus, the distinction between general and domain–specific strategies is a bit fuzzy. The suggestion put forth here, based on the results of the present study, has to do more with the connection between the two rather than the distinction. The suggestion is this: *In order to effectively implement a general strategy one must, in effect, transform it into a domain–specific strategy.*

Consider some of the results discussed above. "Reformulation" is a broadly applicable, general strategy. However, to make it specifically useful, students appear to have drawn upon domain–specific conceptual understanding in order to implement it as "reformulate the group axioms in terms of the operation Δ" or "reformulate $x^2 = e$ as $x = x^{-1}$" or "reformulate 'x in S' as '$xa = ax$'" The actual strategy implemented was extremely domain-specific. As another example, the quite general strategy "introduce notation" was implemented as the domain–specific strategy "let the identity be $e_\Diamond$". Once again, the link back to the conceptual understanding (understanding of different operations generating different structures) transformed the general–but–weak strategy into a powerful domain–specific strategy.

As two final examples, consider two general strategies that students at higher levels used more often: guess and check; and work forward. The data show that upon implementation both processes were used in a very domain specific way. But beyond the empirical evidence, one can see through an a priori analysis that for effective implementation both must be transformed into understanding–based, domain–specific strategies. To guess and check effectively one evidently needs a good guess, which must be generated from conceptual understanding. Similarly, working forward effectively requires a good conceptual map of the specific domain so that one sees where and how to proceed.

Thus, it is suggested that one way to characterize the differences in process use between higher and lower level students is that students at the higher levels were able, based on more developed conceptual schemata, to take a given strategy and implement it in a more powerful domain–specific way.

Another difference between students at different levels can be seen in terms of their control of strategies. This is discussed in the next section.

Metacognition

The key result of this study pertaining to metacognition is the finding that, while there were no differences with respect to strategies available at different levels, there were differences with respect to which strategies were chosen or evoked. That is, there was a difference between higher and lower levels with respect to the control of strategy use — students at higher levels exhibited better control of their process use.

This result is another verification, in a new domain, of a well-documented finding. The interpretation, in the light of linkage to conceptual schemata, is that the relationship between control and level of conceptual understanding is more than just correlational. It is suggested that a problem solver's ability to control process use is based on that problem-solver's conceptual schema.

This interpretation is consistent with the results. For example, higher level students chose to use the process "work forward" more often, while lower level students used "work backward" more often. These strategy choices might reasonably be due to the fact that to work forward effectively requires a complete conceptual model, as discussed above, while working backward is a strategy to choose if you have an incomplete understanding. Thus, the conceptual schema possessed by a student largely determines the strategies chosen.

Another way to consider this issue is in terms of evoking, rather than choosing, strategies. From this perspective, one can view the control of strategy use as being even more closely linked to conceptual schemata. The idea here is that strategies are *evoked*, based on the interaction between the task at hand and the current conceptual schema.

EXPERT/NOVICE

All of the discussion of the results of this study could be in terms of experts and novices, where the upper-level students are experts and the lower-level students are novices. One should keep in mind, however, that, since the levels were defined in terms of conceptual understanding, expertise would be "content–domain expertise," rather than "problem–solving expertise" as put forth by Schoenfeld (1982). In any case, there are two results which have not yet been discussed that have particular relevance to the expert/novice paradigm.

First of all, the fact that almost all students fit into one of the four levels in this study indicates that there are gradations of experts and novices. Secondly, the inconsistent trend across levels with respect to process-use indicates that the evolution from novice to expert may be a rather unstable, irregular, developmental process. Both of these results are consistent with claims made by other researchers in other content domains (e.g., Silver, 1985; Lesh, 1985).

LEVELS MODELS

The levels model most relevant to the present study is the van Hiele model. Hoffer (1983) claims that the van Hiele model, which applies to geometry, can be abstracted and made applicable to all areas of mathematics. However, the present investigator found it very difficult to apply it to group theory. One problem is that the levels tend to become very content-dependent. For example, the levels in this study — syntactic, concrete semantic, and abstract semantic — work well for the explicitly defined content domain of elementary group theory, but it is not the case that any problem in group theory will fit neatly into one of these levels. Furthermore, the levels do not have the recursive structure characteristic of the van Hiele levels. Attempts to find recursive levels in the pilot studies did not meet with success. Another problem is that the middle levels of conceptual understanding seem to be very unstable. Thus they may be hard to characterize. The problem of extending the van Hiele model to other areas of mathematics is an interesting one, and it seems theoretically possible. The questions are: can it be done in practice, and will it be a fruitful endeavor?

TEACHING IMPLICATIONS

There are several teaching implications of this study. Most broadly, the results indicate that our major goal as teachers should be to promote students' acquisition of stable, powerful, accessible conceptual schemata.

In doing this, however, we should not try simply to train novices to behave in the same way as experts. The results of the study show that there are indeed observable differences between experts and novices, but they also indicate that the transition from novice to expert is developmental and unstable. Thus, it is not advisable to try to train novices abruptly in the proof–writing behavior of experts. A vivid example of this is the result reported concerning the process "change plan." Level 3 students (the experts) used "change plan" less than level 0 students (the novices). This certainly does not imply that we should teach novices to change plans less. In fact, the use of "change plan" increased from level 0 to level 1 to level 2.

Thus, rather than try to teach novices the behavior exhibited by experts, we need to find a way to teach novices so that they acquire their own stable and powerful conceptual schemata. This will naturally involve teaching heuristics and metacognition as well as concepts.

Conclusion

The overall conclusion of this study is that we cannot understand proof–writing performance by studying processes, errors, outcomes, and metacognition in isolation. It is not even enough to consider all these variables concurrently. A more holistic approach is needed – one which recognizes that these variables are interdependent and are strongly linked to conceptual schemata (see Figure 1).

Figure 1

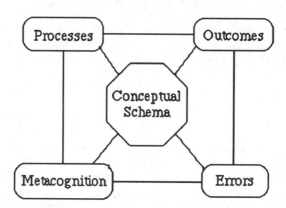

We could even say that processes, metacognition, and misconceptions are actually *part* of one's conceptual schema. This point of view is alluded to by Lesh (1985) when he states that, "processes ... are integral parts of the underlying conceptual models" (p. 324).

Certainly we must deal with many issues in our efforts to gain knowledge and solve problems in mathematics education. These issues range from local ones such as the relationship of certain variables to proof–writing performance, to more global issues such as domain–specific strategies versus general strategies, or heuristics versus content knowledge, or procedural knowledge versus conceptual knowledge, to, finally, the most global issue, which has to do with the interaction of the three fundamental components of education — knower (students and teachers), process of knowing (learning and teaching), and known (content and curricula). The general recommendation offered here is that we must continually strive for a holistic perspective — one which looks for interdependencies and ultimately synthesizes the parts into a wholeness. Such a holistic perspective is suggested as the key to success in both teaching and research.

Norman and Prichard offer a particular set of theoretical frameworks in which to view student thinking in calculus: one practical, based on Cocking and Chipman's conditions for learning; one philosophical, rooted in Bachelard's notion of "epistemological obstacle," specialized to cognition; and one psychological, based on that of the influential Soviet mathematics educator and psychologist Krutetskii. They look at a rich collection of data accumulated in standard calculus and differential equations courses, documented in the form of taped student interviews, and systematically-examined written student work. The role of the frameworks is to classify the observed phenomena, most of which are likely to strike the reader as familiar.

Their taxonomy of cognitive obstacles to learning, at the grossest level, involves a distinction between linguistic/representational phenomena and phenomena associated with the growth of concepts and intuitions. But they go on to examine many subvarieties that occur with regularity among early university students. They confront, but do not resolve, the question of how these are related to prior instruction. It seems apparent that some, based on interference between natural language and mathematical symbolism, are virtually unavoidable, as are those involving perceptual factors associated with graph interpretation. We might say the same about certain tendencies to overgeneralize linearity, for example, which are rooted in the potent human tendency to abstract invariance from similar instances. However, being able to predict difficulty is a most important step in dealing with it.

We suggest that this paper opens a large opportunity both to enrich the set of examples, and to elaborate the distinctions still further. For example, does the well–documented tendency to interpret a graph of some phenomenon with an associated picture (e.g., the picture of a biker pedaling over a hill with a hill–shaped velocity graph — which is almost exactly wrong) have anything to do with the tendency to assume resemblance between a function and its derivative? Are there deeper underlying cognitive factors that undergird seemingly disparate phenomena — beyond those already identified?

This paper presents a very rich set of beginnings, one which the alert reader is likely to capitalize on with little cognitive difficulty.

COGNITIVE OBSTACLES TO THE LEARNING OF CALCULUS: A KRUKETSKIIAN PERSPECTIVE

F. Alexander Norman
University of Texas at San Antonio

Mary Kim Prichard
University of North Carolina at Charlotte

If recent reports from teachers of calculus, mathematics-education researchers, and mathematicians are to be believed, then the learning of calculus in this country is in an abysmal state. Students, of course, pin the blame on instruction. In fact, a number of factors have been identified that affect the apprehension of mathematical concepts in general and calculus in particular. These include social and cultural influences, historical effects, linguistic and representational factors, cognitive development, epistemological perspectives, pedagogical factors, attitudinal and behavioral factors, and others.

Cocking and Chipman (1988) have proposed a simple model categorizing many of these influences on school mathematics learning. This model provides a broad–based perspective on variables relevant to cognitive development and on the influences of both linguistic and non–linguistic factors on mathematical understanding. This model (see Figure 1) presents three general categories: (1) Entry Mastery, (2) Opportunity to Learn, and (3) Student Motivation.

An expanded Input portion of the model describes the kinds of variables included in each of these categories. In the Entry Mastery category, these are the cognitive ability patterns that students bring with them to the learning experience:
(1) Mathematical Concepts, (2) Language Skills, (3) Reading, and (4) Learning Ability.

Figure 1
Influences on learning

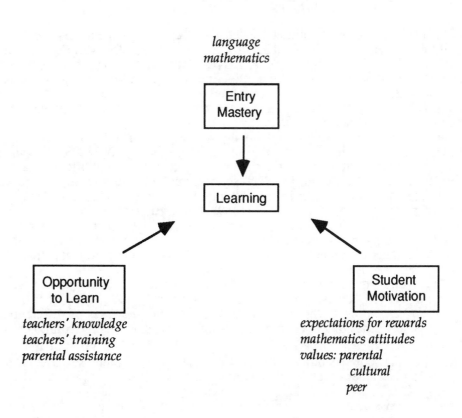

Purpose

Reforms are being called for in both the content and teaching of college–level calculus (Douglas, 1986). Changes in the mathematical skills demanded by society, changes in our student population, and changes in the technology available to both students and teachers are requiring curricular reform at the college level. A report from a workshop on teaching methods in calculus made the following suggestions:

> We need to know more about what students learn in their mathematics classes. A close look at students' work (by means of interviews, videotapes of students working problems, etc.) is often a disturbing, but valuable source of information. More detailed research on students' mathematics learning would be helpful, both to tell us about current difficulties in instruction and to suggest ways that might help us to improve. (Douglas, 1986, xx)

The purpose of this paper is to propose a scheme for understanding students' difficulties in the learning of calculus. We are primarily concerned with the Entry Mastery category of Cocking and Chipman's model, for therein lies the genesis of most of the cognitive obstacles to the learning of calculus. While difficulties might be traced to a student's intellectual deficiencies or to a teacher's incompetence, neither of these is relevant to our discussion. We are concerned with structures that arise "naturally" during an individual's cognitive development and that influence or constrain the further development of mathematical knowledge.

In this paper, we discuss two projects in which we collected data on students' difficulties in applying concepts and procedures to solve a broad range of calculus problems. We used two different methods of data collection. In one study, individual interviews were conducted with students in a first semester calculus class. We also collected homework and tests from students in a first semester calculus class and a differential equations class to analyze their errors and misunderstandings. The tasks used in both settings are fairly typical of the type students encounter in their classes and textbooks.

The theoretical basis for our analysis of students' difficulties lies in two different but related areas of research. Krutetskii (1976) identified characteristics of the problem–solving processes used by students to solve mathematical problems. The three processes that we feel are most closely related to the understanding of the calculus are *flexibility, reversibility,* and *generalization*. The research on cognitive obstacles also provides a framework for looking at students' difficulties in calculus.

KRUTETSKII'S PROBLEM–SOLVING PROCESSES

Krutetskii (1976) identified several abilities related to successful problem solving: flexibility in thinking, reversibility of thought, and an ability to generalize. Most studies in the learning of calculus have focused on students' understanding of particular concepts, such as limit, function, or derivative. Our approach is closer to that taken by Rachlin (1987) in his study of an experimental algebra curriculum. In that project the problem–solving processes of generalization, reversibility, and flexibility guided the selection and creation of tasks that were used in individual student interviews and in the curriculum materials. Here we examine how these processes emerge in the normal conduct of students' work in calculus.

Flexibility. Krutetskii describes flexibility of mental processes as "an ability to switch rapidly from one operation to another, from one train of thought to another" (1976, pp. 222–223). One instance of flexibility in thinking is when students are able to solve the same problem in several different ways. Students who are able to choose from several different problem–solving methods are able to solve a variety of problems posed in many different ways. Another instance involves a student's ability to understand and use multiple representations of a concept (see Kaput, 1987, 1989). In the calculus, the concept of function is central to the understanding of limit, derivative, and integral. Consequently, students who are comfortable with the algebraic formulations and the graphical and numerical representations of functions and who have the facility to make connections between them are better prepared to solve many problems that arise in the study of calculus.

Reversibility. "The reversibility of a mental process here means a reconstruction of its direction in the sense of switching from a direct to a reverse train of thought" (Krutetskii, 1976, p. 287). Reversibility is essential in the complete understanding of the relationship between the processes of differentiation and integration, for example. In many cases the process of integration is introduced as the solution to problems such as "Name a function, F, whose derivative is f." Unfortunately, when rules of integration are introduced, many students do not use reversibility to reason and understand the processes. They tend to view differentiation and integration as rather unrelated processes each of which has its own set of rules to be memorized (e.g., Eisenberg, 1990).

Another example — many students have no difficulty in describing the properties of a given function and its derivative. However, when students are given the properties of a function and its derivative, many are unable to construct a graph of a function that satisfies these properties much less formulate an algebraic expression for such a function.

Generalization. As described by Krutetskii, there are two aspects of generalizability to consider: "(1) a person's ability to see something general and known to him in what is particular and concrete (subsuming a particular case under a known general concept), and (2) the ability to see something general and still unknown to him in what is isolated and particular (to deduce the general from particular cases, to form a concept)" (1976, p. 237). Students learn many rules for differentiating functions in a first–semester calculus class. When given a function to differentiate, students must recognize the function as a particular case to which one of the rules may be applied.

Unfortunately, students are not given the opportunity to apply the second ability very often in their calculus classes. Instructors who feel the need to complete a syllabus tend to tell students the rule or generalization rather than giving them the opportunity to discover it themselves. However, classroom experience has shown that, for example, students are able to generate the power rule for differentiating functions on their own when simply given the opportunity to investigate several cases.

A third aspect of generalization, not touched upon by Krutetskii but which is relevant to the comments above, differentiates among two levels of general rules — rules related to a specific class of functions and general rules for all functions. For example, there is a rule for differentiating functions of the form Ae^{Bx}. There is also a rule for differentiating $f \cdot g$, for any differentiable functions f and g. These two levels of generality are not usually acknowledged by instructors or texts, and certainly most students do not recognize the differences.

COGNITIVE OBSTACLES

As mentioned earlier, there are a number of factors influencing the learning of mathematics. Referring to the model in Figure 1, our concern in this paper is primarily in the Entry Mastery category. This component is particularly relevant since many of the cognitive obstacles we discuss are based in the mathematical and linguistic cognitions of students. Further, as a student begins the construction of new concepts (or the modification of previously constructed ones) in different mathematical contexts, the variety and strength of extant cognitive obstacles influence that student's particular conceptualizations.

The notion of *cognitive* obstacle is closely related to the broader concept of *epistemological* obstacle, described a half–century ago by the French philosopher–scientist Gaston Bachelard (1983/1938). Bachelard perceived the growth of scientific knowledge as constrained, not so much by technical difficulties or the complexities in the analysis of phenomena, but rather by intrinsic, often unrecognized, factors associated with the very process of understanding — factors which are indelibly imbued with the reigning assumptions regarding the nature and growth of scientific knowledge. These constraints, among which he included the influence of natural language, the tendency to generalize, and the reliance on possibly deceptive intuitions, were identified as potential epistemological obstacles to the advancement of scientific thinking. Bachelard's views anticipated those of Kuhn (1970) who, in his book *The Structure of Scientific Revolutions*, points to the sudden ascendance of radically new directions in scientific thought — a breaking and reformulation of scientific paradigms — as the driving force behind scientific revolutions.

While Kuhn and Bachelard spoke primarily of science in its most general form, the history of mathematics is replete with examples of revolutionary thinking that refused to be bound by the constraints of the epistemology in vogue. The introduction of negative integers, Descartes' analytic geometry, the rise of non–Euclidean geometries, the current focus on chaotic systems and computer–generated proofs are all instances in which some powerful epistemological obstacle was overcome leading to revolutionary advances in mathematical thought.

In contrast to epistemological obstacles, which exist within a context of scientific thought in general, *cognitive obstacles* are idiosyncratic to an individual's own learning experiences. As Herscovics (1989) puts it:

> ... just as the development of science is strewn with *epistemological obstacles*, the acquisition of new conceptual schemata by the learner is strewn with *cognitive obstacles* (p. 61).

Likewise, just as epistemological obstacles can constrain the growth of science, cognitive obstacles can constrain the development of a mathematical concept. Without delving too deeply into the psychological basis for this, we note that a constructivist view of concept development provides a theoretical framework in which to view cognitive obstacles. Within this perspective the nature of an individual's conceptualization is influenced by cognitive structures and processes already extant. For example, to use Piagetian terminology, the accommodation of a new concept requires a reorganization of cognitive structures, an action that may be not be so easily or appropriately done. In fact, Herscovics speaks of "their [cognitive structures] becoming cognitive obstacles in the construction of new structures." (p. 62)

Note that the existence of cognitive obstacles is in no way to be construed as pejorative. In fact, many of these obstacles are likely a very natural part of cognitive development. The extent of the influence of the obstacles is what is problematic. Thus, understanding the origin of a particular obstacle, how it relates to an individual's unique mathematical experiences, and how it might be overcome are questions that are useful to examine. As an illustration of these points, consider the difficulty that many calculus students have in understanding the object nature of a function (see, for example, Sfard, 1990). The difficulty for these students in this situation lies in their prior conceptualization of a function as an action or process. This process notion is, in a sense, antithetical to the view of a function as an object, and thus is a strong cognitive obstacle to perceiving it as such. In fact, a number of researchers have focused their attention on the development of functional understanding by looking at how students make the cyclical transitions from process to object notions (Dubinsky & Harel, 1990; Breidenbach, Dubinsky, Hawks, & Nichols, 1992).

A STUDY OF PROBLEM SOLVING ABILITIES

Mathematicians and mathematics educators agree that one of the goals of mathematics instruction is to help students apply concepts and procedures to solve a broad range of problems. Most calculus students seem unable to solve calculus problems that are presented in a form different from that discussed in class or presented in their texts. They approach the learning of problem–solving as they would the learning of procedures. Many times they look for rules instead of patterns; they memorize rather than analyze. The goal of this study is to determine to what extent students are able to demonstrate reversibility and flexibility of reasoning processes and to generalize their reasoning processes when solving calculus problems.

To study the process of students' thinking rather than simply the products of their thinking activities, four clinical interviews were conducted with each of six students enrolled in a Calculus I class. During the interviews students were videotaped or audiotaped as they solved a series of calculus problems. This approach provided the detailed trace of students' problem–solving behaviors that was needed for this study. As an illustration of the data collected from these interviews, we consider the case of Paige, one of the students participating in the study. Paige was one of the students who volunteered to be interviewed. In high school, Paige had made A's in 10th grade Geometry, B's in 11th grade Algebra II, and B's in Trigonometry in the 12th grade. Paige was a second–semester sophomore when she enrolled in the Calculus I course. She had already had Pre–Calculus at UNCC twice — the first time she made a D, then repeated the course and made a B. She had also taken a Business Calculus course. While Paige was not a very good student, her misunderstandings and difficulties are fairly typical of many calculus students.

ABILITY TO GENERALIZE

The following are typical tasks focusing on the recognition of a particular case of a known concept or rule:

GENERAL → PARTICULAR

If $f(x) = cx^n$, then $f'(x) = cnx^{n-1}$.

1) $f(x) = -2x^5$, $f'(x) =$ _____ .

2) $f(x) = x^{-2}/3$, $f'(x) =$ _____ .

3) $f(x) = \frac{1}{2} \cdot x \sqrt{\frac{3}{2}}$, $f'(x) =$ _____ .

This last task was encountered by Paige while solving a related rates problem. Rather than recognizing this expression as a special case of cx^n, to which the simple power rule could be applied, she focused on the product of the two factors, $\frac{1}{2}x$ and $\sqrt{\frac{3}{2}}$, and applied the product rule. (In a similar problem, another student used the quotient rule to find the derivative of $\dfrac{\sqrt{3}x^2}{2}$.) Most calculus instructors would assume that the exponent rule is mastered by the majority of their students. However, many students do not recognize that they can apply this rule to functions that involve numbers other than rational numbers.

FLEXIBILITY IN THINKING

    I. Another task involved solving the same problem in several different ways.

$$f(x)=\begin{cases} x^2+1, & \text{if } x<-2 \\ 3x+2, & \text{if } -2\le x<1 \\ x+1, & \text{if } x=1 \\ 6-|x-2|, & \text{if } x>1 \end{cases} \quad \textit{(the appropriate graph was given)}$$

1. Given the function f expressed algebraically as well as graphically, what is the limit of $f(x)$ as x approaches 0? As x approaches 1?

2. Can you find the limit of $f(x)$ as x approaches 0 in some other way? How about as x approaches 1?

    When Paige answered limit questions about this particular function, she used the graph. She recognized that $\lim_{x\to 0} f(x)=f(0)=2$. She also recognized that $\lim_{x\to 1} f(x)=5$. However, she was unable to use the algebraic formula for $f(x)$ to determine a limit without considerable prompting.

    When given a function expressed in its algebraic form (but no graph) and asked to find $\lim_{x\to 1} f(x)$ Paige automatically evaluated $f(1)$. When f was a piecewise–defined function, she was unable to evaluate $f(1)$. When asked to think of another method to solve the problem, she did not suggest sketching a graph to confirm or determine her answers.

    II. Flexibility across problems is the degree to which a successful solution process on a previous problem fixes a student's approach to a subsequent problem (Rachlin, 1987). A good problem-solver knows when to "fix" and when not to. This process is closely related to the process of generalizing in that a problem solver may recognize that a new problem situation is similar to one recently encountered and that the same problem–solving approach would be helpful for the new situation. On the other hand, a good problem-solver also recognizes when a new problem–solving approach is more effective for a particular problem.

    Paige was given only the graph of a quadratic function [$f(x)=(x-1)^2/16$] and asked to find the following:

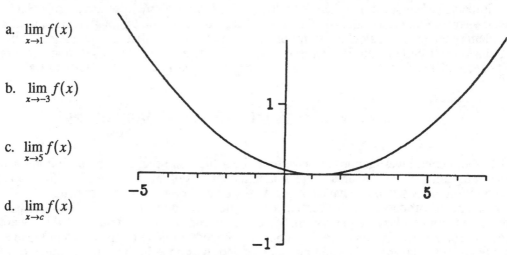

    a. $\lim_{x\to 1} f(x)$

    b. $\lim_{x\to -3} f(x)$

    c. $\lim_{x\to 5} f(x)$

    d. $\lim_{x\to c} f(x)$

    In questions a, b, and c, Paige located each point to correctly identify the limits. When she attempted d, she had no idea of how to describe what the limit should be. She had "fixed" on the method of locating points and was unable to use another strategy for solving the problem. It is also possible that her expectation of what her answer should be did not allow her to describe the limit in a way that seemed to her too vague — namely, $f(c)$.

    Paige's responses to these tasks and her inability to demonstrate flexibility in her thinking is related to her difficulty in working with the multiple representations of functions. Other students who understood and used the algebraic and graphical representations of functions were not challenged by these tasks. They were solved easily and quickly.

REVERSIBILITY OF THOUGHT

I.   Sketch a possible graph for a function f such that:
      a. f is everywhere continuous
      b. $f(-3)=1$, $f(0)=2$
      c. $f'(x)<0$ when $x<0$.

Like many of our calculus students, Paige was not very successful with problems of this type. This could be due in part to her lack of experience with problems of any type that require reversibility of thought. Paige hesitated before plotting the points $(-3,1)$, and $(-0,2)$. Then she had no idea about what steps to take next. She could interpret the third condition in terms of the slope of tangent lines, but she did not know how to use that information to sketch a graph.

II.   Find a function whose derivative is $3x^2$.

Paige had difficulty answering a question as seemingly simple as this one. Her understanding of the relationship between the derivative and integral was weak. When Paige began to study integrals and basic techniques of integration (power rule and the method of substitution), she approached differentiation and integration as two different sets of rules to memorize rather than as reversible processes. When the same task was presented in the form $\int 3x^2\,dx =$ _____, she easily used the power rule for integrals to find x^3+c .

REMARKS

The three abilities — generalizing, flexibility of thought, and reversibility of thinking processes — provide a framework for studying and describing students' understanding of calculus and their performance on calculus problems. While these abilities are very important in successfully solving problems in calculus, they do not receive explicit attention by classroom teachers and curriculum developers. In particular, relatively few problems in a standard textbook show different methods for solving calculus problems. Paige's inability to use these processes hampered her progress in the study of calculus. Other students who demonstrated greater facility with these three processes tended to have more success in the course.

Students' understanding of and facility with the multiple representation of calculus concepts and processes — symbolic, graphical, and numerical — are essential for a thorough understanding of the calculus and problems that require calculus. Many of the current projects that are revising calculus curricula and teaching have identified the multiple representation of concepts and processes as a major component of their work. [See, e.g., Dick & Patton, 1991 (Oregon State Calculus Project) and the Harvard Calculus Project.]

AN ANALYSIS OF COGNITIVE OBSTACLES

In the second investigation, we have been, over the past several years, examining and cataloging the work of our calculus students in an attempt to identify sources of misunderstanding. In an attempt to keep the setting as natural as possible, we have conducted the study in a non–intrusive manner by simply examining the work students produced in the context of the normal conduct of their classes. (For example, in a differential equations class taught by one of us, all errors on homework and tests were recorded and analyzed.) In some cases, during review sessions, we probed to find out why a student made a particular error or exhibited a particular misunderstanding, but this, too, was done in the natural course of instruction. Data have been collected from several classes, but those from two classes (a first–semester calculus class and a differential equations class) have been completely and systematically analyzed. The comments that follow are in regard to data from these two classes.

The cognitive processes of flexibility of thought, generalization, and reversibility provide a framework for analyzing and classifying specific cognitive obstacles identified during our investigation. In the first study, typified in the responses of Paige, we see a number of examples that provide insight into common cognitive obstacles from the perspective of this Krutetskiian framework. The results of the second study, as well as the work of other researchers, provide additional instances of cognitive obstacles that appear to be closely tied to these three processes.

Results of the first investigation indicate that while many of the errors students made were, of course, arithmetical or procedural in nature, a significant number[2] of errors appear to be tied to specific, sometimes well–known, cognitive obstacles. In many cases, the obstacles identified are simply analogues of those that have been identified in the extensive research on algebra learning [see Kieran (1989) and Hersovics (1989) for good reviews of that research]. In fact, difficulties with algebra itself seem to be frequent impediments to students' success in their calculus courses.

We have identified two general types of cognitive obstacles, which parallel on a local scale, epistemological obstacles identified by Bachelard. These obstacles are associated with (1) natural and mathematical language, and (2) intuitions about mathematical representations and specific mathematical concepts, such as limit or continuity. Furthermore, the cognitive mechanisms driving these obstacles can be tied to Krutetskii's processes of flexibility, generalization, and reversibility.

LINGUISTIC/REPRESENTATIONAL

Mathematics has, or perhaps is, a very peculiar and sometimes abstruse language. Mathematical concepts often defy description in natural language so that students who do not fully understand mathematical syntax, conventions and, even, mathematical idioms, may find it difficult to apprehend fully the concepts they are trying to learn. Students sometimes impose a natural language grammar on mathematical situations (a tendency to generalize inappropriately) or make semantic judgments based on a misinterpretation of syntax.

The symbolism of written mathematical language also provides an environment conducive to the emergence of cognitive obstacles. For example, Skemp (1982) has pointed out that there are many more mathematical concepts, operations, and processes than there are symbols used to denote them. This can lead to various difficulties with interpreting mathematical symbolism, among which those related to synonomy (different symbolic expressions having equivalent meanings) and homonomy (different meanings for identical expressions) are not uncommon (Adda, 1982). Thus, understanding the relevant context is essential for a reasonable interpretation of the symbolism.[3] The way individual and groups of symbols are perceived might also have an effect on interpretation. Higginson (1982) and others have suggested reformulations of mathematical symbolism to take into account the perceptual aspect of symbol interpretation — for example, developing operator symbols that visibly reflect the underlying action of the operation being used.

In the examples that follow, we discuss cognitive obstacles related to concatenation, the name–process dilemma, unitizing, operation transference, generalized distributivity, and natural language. Many of the problems delineated here you have probably noted among your own students.

Example 1. Consider the problem of *concatenation*. Difficulties with concatenation have been identified among beginning algebra students (Collis, 1974). For instance, these students, who have had much experience with juxtaposed numerals, easily recognize that the meanings of "32" and "23" are quite different. However, when this syntactic interpretation is maintained, their interpretation of the algebraic expression "$3x$" becomes 'thirtysomething' and the expressions xy and yx appear to be different. In this case, their arithmetic interpretations are a cognitive obstacle to an appropriate conceptualization of algebraic juxtaposition. Eventually, for most students, this obstacle is overcome and a multiplicative interpretation of such algebraic expressions is constructed.

However, this new multiplicative interpretation can become problematic when functional notation appears — in that context the juxtapositions $y(x)$ or $\sin(\cos x)$ should not be considered multiplicatively. As an example from calculus that points to problems with concatenation, we have noted instances in which students reverted to a multiplicative interpretation of expressions by performing the transformation $\int fg \to f\int g$.

A typical instance of this sort is the student who cleverly avoided that annoying integration by parts with the following technique:

$$\int x \cdot \ln(x)\, dx = x \int \ln(x)\, dx = x\left(\frac{1}{x}\right) + C = 1 + C.$$

---

[2]Approximately 10,400 errors were analyzed of which about 22% were not simply arithmetical or procedural.

[3]This is an affliction of both oral and written natural language as well. For example, responding for the first time to a comment that he had bare feet, my 3 year old said, "Not bear feet. I got people feet."

This is a clear instance of generalizing a rule from a specific, and appropriate, context within a context for which the rule is not legitimate. To understand the source of this error, we note that it is not the overgeneralization that is the cognitive obstacle, but rather the very tendency to generalize. The overgeneralization is simply a natural manifestation of this obstacle. Students must be able to generalize to make sense of mathematics; yet, as this example illustrates, the process of generalization can be an obstacle to the correct interpretation of mathematical language.

Example 2. Davis (1975) has described a *name–process* dilemma that faces many beginning algebra students. Almost invariably, during their arithmetic training, expressions involving operation signs, such as $3+2$, were transformed into the closed form of a single numeral. When faced with algebraic expressions such as $2+x$,, some students do not accept this apparently unclosed form and insist on converting it into a single expression (such as $2x$). Of course, we know that the expression already can be treated as a unit if we wish, but this tendency to view the expression as something which must be processed seems powerful in many students.

An analogue of this appears in certain calculus situations as well. For example, when asked to find $F(a)$ where $F(x)$ was defined as $F(x)=\int_a^x f(t)\,dt$, several students in one class declared that such a thing was impossible without first knowing f. Then, when supplied with a convenient $f(t)=t^2$, most of these proceeded to do essentially the following:

$$F(x)=\int_a^x t^2\,dt=\tfrac{1}{3}x^3-\tfrac{1}{3}a^3,\text{ so } F(a)=\tfrac{1}{3}a^3-\tfrac{1}{3}a^3=0.$$

This and similar examples suggest that some students seem impelled to *do* something whenever they see the integral symbol. We have found it the exception, rather than the rule, when students, even good students, really think about a definite integral as a number, rather than as an operation that outputs a number. This perspective of the definite integral as requiring an action, rather than as representing an instance of a process, exemplifies the name–process dilemma in a calculus context. The difficulty here might arise in several ways. Certainly it reflects the action/process/object interpretation of concept development (Breidenbach et al., in press) and is consonant with the distinction Tall (1991) has made between a process and its product.

We also believe that the Krutetskiian construct of reversibility of thought has relevance to some instances of the name–process dilemma. For the most part, students' only experience with integrals (definite or otherwise) is the conversion of those integrals into numerical values or more familiar functional forms. This conversion often is understood only unidirectionally — i.e. $\int x^2\,dx$ produces $\tfrac{1}{3}x^3$, but the latter expression is never converted into, nor is it thought of in terms of, the former.[4] Students who are unable to reverse their thinking about the relationship of these expressions are unlikely to recognize their essential equivalence.

Example 3. A problem that has been consistently noted among some calculus students in our investigation is a difficulty with *unitizing* variable complexes. Norman (1986, 1987) has described several different unitizing strategies that students exhibit in dealing with variable complexes (e.g., when factoring the expression $3(y+2)^2-2(y+2)-1$, one might perceive the complex $y+2$ as a unit and use this to factor the reconceptualized expression $3U^2-2U-1$).

There are many instances in calculus in which students must recognize the unitary nature of variable complexes or create their own complexes. Function compositions, the chain rule for derivatives, and integration techniques that involve substitutions are all areas in which a high level of unitizing skill is essential. The disinclination to unitize is certainly a cognitive obstacle to a complete understanding of these.

Example 4. Operation transference is a form of generalization. It refers to the tendency to transfer rules related to an operation in one context to a completely different context. Sometimes the transfer is legitimate; sometimes not. For example, $A+B=B+A$ where A and B are real numbers or real $n\times m$ matrices. But, while we know that $AB=BA$ for real numbers, this is not necessarily so for matrices.

These types of generalizations have been identified among algebra students (Matz, 1979), but are prevalent among calculus students as well. Students enter calculus with the exponent rule $(xy)^n=x^n y^n$ well in hand. We have noted that some students have a tendency to apply this rule inappropriately with derivatives, writing, for instance,

$$(fg)'=f'g' \text{ and } (fg)''=f''g''.$$

---

[4]This is analogous to the arithmetic situation in which a student can process the addition 3+5 to get 8, but cannot "unpack" 8 as 3+5.

For these students, it must seem very odd indeed that

$$(x+y)^2 = x^2 + 2xy + y^2 \quad \text{and} \quad (xy)^2 = x^2 y^2 ,$$

while

$$(f+g)'' = f'' + g'' \quad \text{and} \quad (fg)'' = f'' + 2f'g' + g''.$$

A similar problem that was noted frequently among several students in a differential equations class is exemplified in the following problem solution:

$$y'' + 2y' = 0 \quad \Rightarrow \quad y'(y' + 2) = 0$$
$$\Rightarrow \quad y = c, y = -2x + c$$

Example 5. Another difficulty with which we are all too familiar is that of *overgeneralized distributivity*, specifically of the form

$$F(a*b) = F(a)*F(b)$$

in which $*$ is a binary operation and F a function that is not linear with respect to $*$ (see Matz, 1983). While we seem to have trained many of our students not to make the error of writing $(x+y)^n = x^n + y^n$ (assuming $xy \neq 0$), this does not seem to inhibit students from making equally egregious distributivity errors in other contexts. For example, $\ln(x+y) = \ln x + \ln y$ (or something equivalent) was noted to occur at least once among nearly 80% of the Calculus I students. Even among our more advanced students we have often seen expressions like

$$e^{\int p(x)\,dx+C} \rightarrow e^{\int p(x)\,dx} + e^C$$

Why are these kinds of errors so durable? Actually, it is not so surprising. Mathematical rules which are not applicable in syntactically isomorphic situations can be powerful cognitive obstacles. There are many instances in certain contexts in which distributivity *is* appropriate — linear transformations, homomorphisms, differential and integral operators, even the definitions of addition and multiplication of functions has a superficial distributivity — $(p+q)(x) = p(x) + q(x)$ and $(pq)(x) = p(x)q(x)$. Clearly the problem here lies in an interpretation of structure. Different mathematical contexts can require different interpretations of structurally isomorphic symbolic expressions. Since students can perceive onlythe symbols that appear on paper or the chalkboard or the computer screen, those who do not fully understand the underlying mathematical context are at great risk to make such errors of structural misinterpretation.

Example 6. The following provides an example of how *natural language* can sometimes place subtle constraints on appropriate interpretation of mathematical problems. In this problem the critical phrase is italicized and the beginning of a typical solution (given by 4 of 27 students) is shown.

> A half–filled reservoir containing .03% fluoride is being filled with a solution of .01% fluoride at a constant rate while the mixed solution is being simultaneously drained. If, during each second, *twice as many gallons are coming in as are going out*, write an equation …

Solution:

$$\left. \begin{array}{l} I = \text{amt. input} \\ O = \text{amt. output} \end{array} \right\} \rightarrow 2I = 0$$

$$\vdots$$

This variant of the well–known Student–Professors problem (Clement, Lochhead, & Monk, 1981) suggests that natural-language problems of the simplest kind can lead to difficulties. The difficulty here, though, is primarily an algebraic one, rather than a calculus one. Nevertheless, mathematics instructors at all levels need to be aware that misreadings and unexpected interpretations of natural language do occur.

INTUITIVE FACTORS

It is a nearly universal characteristic of the human mind to create or attach meaning to a wide variety of phenomena. While attempting to understand mathematical concepts, students construct their own idiosyncratic meanings for these concepts. In many cases, these constructs are very nearly what we, as mathematicians, would want. However, in other instances, a concept that a student is trying to understand may not be so easily conceptualized. In such cases, the student may rely on intuition about the situation in order to attach meaning to the concept in question. While this is a natural mechanism for dealing with phenomena, a mechanism that can be applied very productively, often students' intuitions about mathematical phenomena tend to result in misconceptions. For instance, some of the examples in the previous section may reflect some of the influence of our students' intuitions about the mathematical "grammar" of symbolic expressions.

There is a growing body of research on students' understanding of calculus concepts [see Graham & Ferrini–Mundy (1989) for a good review of recent research]. Many of these studies have pointed to deceptive intuitions about concepts such as function, continuity, derivative, and integral — intuitions that can act as cognitive obstacles to the development of appropriate understanding. In the following paragraphs, we discuss several of these essential calculus concepts and some of the faulty or incomplete intuitions exhibited by students that can misdirect their conceptualizations.

Function. For the most part, students perceive functions in one of two representations — as symbolic equations or as graphs. However, the connection between the two representations is tenuous (e.g., Wagner, Rachlin, & Jensen, 1984). For some students a graph without an explicit formula is meaningless (Dreyfus & Eisenberg, 1982). Many students' intuitive view of functions is static (i.e., they think about one point at a time) (Monk, 1987). Such perspectives must constrain a student's understanding of the calculus processes that evoke a sense of motion along a curve —for example, the limit of a function or the limiting secant interpretation of tangent line (Tall, 1987; Dreyfus & Eisenberg, 1983; Kaput, 1979).

When the functionality of graphs becomes a consideration, atomistic and other discontinuous functions do not seem to fit students' intuitions about the graphical nature of functions. This difficulty has been noted by other researchers (e.g., Dreyfus & Eisenburg, 1983; Graham & Ferrini–Mundy, 1989) and was evident among some of the students in this study. It appears that students' intuitive notion of the graph of a function is that of a continuous curve. In addition to the expectation of continuity for functionality, some students even require smoothness. We noted instances (See Figure 2) in which functions with "pointy" graphs were commonly identified as non–functions.

Figure 2
Graphs identified as non–functions.

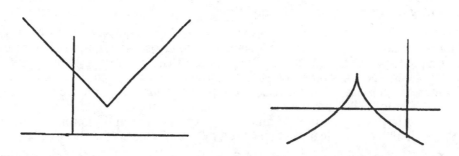

Graham and Ferrini–Mundy (1989) have pointed out that students' intuitions about functions are often limited to what they have explicitly seen in class or their texts. For example, they report that, when a student was asked whether the graph in Figure 3 was a function, the response was, "I think so, yes … it looks similar to a graph that we've been doing in class lately …" (p. 2). Such an *expectation of familiarity*, as the researchers term it, can become an obstacle to students' deepening their understanding of function.

Limits. Students' intuitive notions of limits are quite varied. Some, as illustrated by the dialogue with Paige in the previous study, have difficulty associating the limit of a function at a point with the graph of the function (see also, Graham & Ferrini–Mundy, 1989). In fact, for Paige, a limit is simply the result of a substitution in an equation. Williams (1991) has reported a study of different models of limit among calculus students. This diversity of intuitions — for example, viewing limit as an unreachable value, as an approximation, or in dynamic contexts — appears analogous to the situation in the mathematical community prior to the rigorous formalization of the notion by Cauchy. We wonder why, if it took mathematicians such a long time to formalize the notion of limit, we should expect students to understand adequately the rather unmotivated formalized version presented in calculus courses — and in one class period at that! Students' various intuitions regarding the limit concept can present cognitive obstacles not only to a deeper understanding of the limit concept, but to the understanding of other calculus concepts as well.

Figure 3

A function?

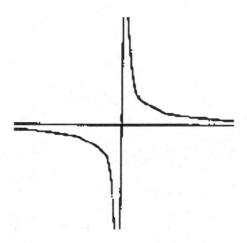

Derivatives. Regarding students' understanding of derivatives, the results of our studies confirm those of Orton (1983), Graham & Ferrini–Mundy (1989), and others. While students are fairly successful at performing certain typical calculus routines, such as differentiating polynomials, their intuitive notions of the derivative of a function seen not nearly so strong. In particular, some of our Calculus I students had difficulty seem reconciling the definition of a derivative of a function at a point with the slope of the tangent line at the point. As has been noted by others, many students have difficulty in seeing that the tangent line is, in fact, the limiting line for a sequence of secant lines (see, e.g., Dubinsky, 1989). Given the graph of an unspecified, but differentiable, function, nearly all of the Calculus I students exhibited difficulty in constructing a graph of its derivative and second derivative. Most of these students' intuitive notions of derivative centered around symbolic differentiation (i.e., most had no graphical intuition); consequently, this limited intuition proved to be a cognitive obstacle to a comprehensive understanding of derivative.

Integration. Geometric intuitions that we consider to be cognitive obstacles to the understanding of integration include those related to notions of area and change of area. These appeared most often as students tried to understand the limiting-sum definition of the definite integral. For example, one intuition exhibited by some students was that the more approximating rectangles one had, the greater the area —regardless of the particular function or whether lower or upper limits were being examined. A second intuition, which is quite natural but interferes with the computation of integrals, is the notion (or lack thereof) of "negative" areas. In certain situations, we noted that students also seemed to associate in a peculiar way the value of an integral (or derivative) with a value of the function. For example, several students indicated that the slope of the curve in Figure 4a was greater at point B than point A because the function value at B was greater. Similarly, another student indicated that the area under the curve — and, hence, the integral $\int_0^x \phi(t)\, dt$ — in 4b decreased as x increased because the "graph is getting smaller ... closer to the axis." [This phenomenon has been noted by Eisenberg (1990), as well.]

Figure 4

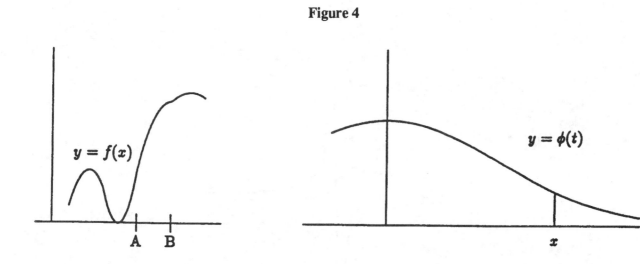

(a) Which is greater —
$f'(A)$ or $f'(B)$?

(b) How is the area under the curve changing as x moves to the right?

Physical phenomena. Students' intuitions about the physical world — spatial sense, motion, rates of change — influence their understanding and interpretation of many of the "applied" problems with which they deal in their calculus courses. We have often noted that students exhibit at best only vague notions of the physical situations of typical word problems. For many it is a challenge simply to translate the problem (even without accompanying understanding) into symbolic form. This latter problem is merely a technical difficulty, however. Some students bring with them intuitions about the physical world that make it difficult to understand what is really going on in such problems. For example, one student had difficulty accepting a problem's requirement that water drain from a conical reservoir at a constant rate. He reasoned (from his experience) that, if a tank were full, then, because of the increased weight of the liquid above the drain, the water would be expelled with greater force (hence volume and rate, as well) than if it were nearly empty. As a second example, several of our students exhibited the same intuitions about the nature of "sliding ladders" reported in detail by Monk (1990). For instance, in the case where the bottom of a ladder is being pulled away from the wall at a constant rate, some students feel that the top must be sliding down the wall at the same constant rate as well. Others have indicated that it must move faster at the bottom than at the top *because falling things move faster and faster as they fall.* Neither intuition is valid for sliding ladders in this situation — even though the latter assumption turns out to reflect reasonably the perceived behavior of the ladder. These examples point out that intuitions about physical situations can be cognitive obstacles to a full understanding of the situation and, therefore, to a correct mathematical interpretation of that situation.

CONCLUDING REMARKS

It should be made clear that the *particular* cognitive obstacles identified are very much tied to the current formulation of mathematical instruction in general and the calculus in particular. More broadly, as we mentioned early on, the factors influencing learning are extensive and varied. However, in no way should they be construed as representing a fixed and permanent set of influences. Instructional methods change; advances in technology open up new ways of representing mathematics; attitudes and beliefs about mathematics change; emphases are refocused. These sorts of influences will inevitably affect the nature of mathematics and mathematics-learning. We determined that cognitive obstacles to the learning of calculus arise in at least two different ways — one related to linguistic/representational aspects and the other related to intuitions. Regardless of the evolution of mathematics and mathematics-learning, it is difficult to imagine that these two aspects, whatever their future formulation, would not continue to be primary influences on the genesis of cognitive obstacles.

Consider, for a moment, the current conditions as an exemplar of linguistic/representational aspects and intuitions as sources for cognitive obstacles to the learning of the calculus. Given that so many of our algebra and calculus courses are immersed in symbolic manipulation, often at the expense of understanding, it is not surprising that linguistic/representational factors give rise to cognitive obstacles. Also, since learners basically *want* to understand — or at least make sense of — what they are being asked to learn, the intuitions that students bring to bear on the concepts of calculus often become obstacles to the appropriate construction of those concepts. We propose that a potentially useful framework in which to embed considerations of cognitive obstacles lies in the cognitive processes of reversibility, flexibility, and generalization.

Aspects related to these investigations that we have not considered, but which are clearly worthy of discussion, are those which touch on the role of external influences on cognitive obstacles. In particular, one aspect that needs to be examined is that of *pedagogical obstacles*. Like their cognitive and epistemological counterparts, such obstacles present constraints, often unforeseen, to the further construction of knowledge. Pedagogical obstacles may act as precursors to certain cognitive obstacles.

As an illustration, consider the traditional presentation of mathematics in our texts and classrooms. This presentation generally tacitly assumes that a formal depiction of a concept (via a mathematically precise description and a few minimal examples) is sufficient for understanding. Students begin to view the essence of mathematics as a series of formalized routines. This particular perspective of mathematics may well constrain students from the investigative activities that are needed to develop a critical understanding of the relevant mathematical concepts.

The time is ripe for reform of the mathematics curricula and the teaching of mathematics, particularly in the critical area of the calculus. Such reforms, though, must be based on an understanding of the nature of students' construction of mathematical knowledge and the variety of factors influencing that construction. This is essential knowledge for mathematics-educators. Perhaps the Krutetskiian framework described here and the discussion of cognitive obstacles will prove useful to those mathematicians and educators who wish to play a role in the reformulation of mathematics-teaching and — more importantly —learning.

This study, labeled as preliminary, probes ground that is new in the sense of being newly subject to systematic inquiry, but is as old as human experience itself. We all know emotion when we experience it, and we all can recognize it in the outward demonstrations of another. And somehow, even that most cognitive of activities, mathematical problem–solving, is deeply enmeshed with emotions of many types and hues. While one might (emotively) yearn for certainty and precision in research, sometimes it is impossible and inappropriate. Fine–grained statistical data on attitudes towards mathematics, on confidence or anxiety, or even on aesthetic judgment of some piece of mathematics based on survey or other data-gathering instruments, may not be quite appropriate in this case, where attention is on the role of emotion in the *process* of problem–solving and where the phenomena to be studied require further identification and clarification.

The author and colleagues examined emotion occurring during paired problem–solving sessions involving several dozen problem-solvers — about 30 experts, ranging from reasonably mathematically-experienced teachers and users of mathematics to professional mathematicians at the faculty level — and a like number of novices, including undergraduates in low and intermediate math courses, as well as a dozen math majors. Data on experts was based on self–reports (concurrent and retrospective) as well as observer notes taken by both the observer–half of the problem–solving pair and the author or her colleagues. Novice data is described later.

The role of aesthetic considerations seems to be central among experts. Is aesthetic judgment behind emotional response, or is it the other way around? How are they related? Solvers seemed quite tightly constrained by aesthetic judgments of the match between solution method and problem level, and by their perception of economic and elegant solutions versus. labored case analyses, for example.

An interesting cyclical variation in attention appeared during problem–solving, where the solver would alternate between intense concentration on the problem, superficial or routine activity (e.g., rewriting the problem statement, cleaning up equations, et cetera), and distracted attention to the external environment. Rosamond also found that imposing time limitations seemed to devalue the problem–solving strategies — solvers felt constrained to be more systematic and less exploratory than they preferred, and less invested in the particular methods they used. Interestingly, five of eight women, but only one of 23 men mentioned the usefulness of mathematics in their discussion. One of the women, a physician, found mathematical problem–solving very unrewarding, almost frivolous, because of its seeming inutility, despite her considerable ability. Perhaps her judgment and feeling were based on a comparison with the problems she solved in her daily work.

Data gathering with the 32 undergraduates was more structured — alternating paired problem–solving and taped group–debriefing discussion sessions. While obviously these students had fewer intellectual resources at their disposal, and the non–mathematics majors among them showed considerable fear, anxiety, and tendency to frustration, they shared some of the same aesthetically based affect that was seen among the experts. The mathematics majors especially showed some of the same cyclical behavior, and follow–up interviews after a course that emphasized paired problem–solving revealed the same consistent tendency that others have reported for the students to internalize the monitoring and reflective role of the partner, which positively affected their own problem–solving behavior when working alone.

The bulk of this paper amounts to identifying anecdotally phenomena that are worthy of subsequent inquiry of a more systematic, although not necessarily quantitative nature. One would hope to see more definitively operationalized characterizations of the cyclical attention phenomena, for example, as well as further exploration of relations between aesthetic and affective factors during the problem–solving process. But we see as essential the focus of such inquiry on the processes themselves, especially as the processes themselves vary — across populations and across problems. It seems that only through such variation are we to have a chance at identifying underlying invariants, especially those that take the form of critical cognitive/affective relationships.

The Role of Emotion: Expert and Novice Mathematical Problem-Solving

A Preliminary Study

Frances A. Rosamond
Department of Mathematics
National University
4125 Camino del Rio South
San Diego, CA 92108

In *A Call for Change: Recommendation for the Mathematical Preparation of Teachers of Mathematics* (Leitzel, 1990), the Mathematical Association of America joins the National Council of Teachers of Mathematics (1989, 1991), the Mathematical Sciences Education Board (1990), Sigma XI (1989), and other professional organizations (Kenschaft, 1990) in including the role of affect or emotion as a construct deserving of the attention of teachers and researchers. Of concern is how a positive attitude toward mathematics can be developed, what techniques can be employed to foster the joy of learning, and how to develop an environment that is perceived as encouraging rather than hostile or overly snobbish and competitive (Buerk, 1982, 1988; Rosamond, 1984). An additional goal is to find ways to involve students more actively in their own learning.

During the last 15 years, considerable progress has been made in understanding the importance of affective variables in relation to mathematics-learning. The initial motivation for most of the work, begun in the mid–seventies, was the desire to understand variables related to gender differences (Fox *et al.*, 1977). By the mid–eighties, isolated mathematicians throughout the world were convinced emotion was a powerful factor in mathematics-learning for all students. In 1985 and 1986, the Canadian Mathematics Education Study Group (CMESG) invited Poland and Rosamond to offer Study Groups to explore the role of feelings in learning mathematics and techniques in organizing classrooms (Rosamond, 1985). In 1986 and 1987 in San Diego, McLeod (1989) arranged a conference to discuss mathematics problem–solving in relation to the theoretical framework of emotion developed by George Mandler (1975).

These discussions revealed that beliefs, attitudes, and values are elements of what has generally been called the affective domain and that they each have an impact on cognition. Mandler prefers to use the term "emotion" to refer to the intense "hot" passions, such as fear, joy, or anger, and which require some kind of visceral arousal and which may be accessed very quickly (Mandler, 1989). The theories of Mandler, Lazarus (1980, 1986), and other information-processing constructivists are attractive to educators because they admit mediating or coping mechanisms that allow evaluation of emotion and cognitive realignment with one's goal. These theories suggest that emotions can be predicted and used to improve mathematics learning (perhaps) as part of what Skemp defined as reflective intelligence: the ability to make one's own mental processes the object of conscious observation and to change these intentionally from a present state to a goal state.

In an attempt to begin to describe the types of emotional reactions that occur during mathematical problem–solving and to identify ways in which emotion inhibits or enhances learning, a two–part research project was designed in which mathematics experts and then novices engaged in observer/solver, paired problem–solving, exercises. This chapter will describe the exercises and results, make some comparisons between the experts and novices, and render some claims about the value of the exercises.

Positive Emotions and Mathematics Problem–Solving

It was hoped that positive emotions that contributed to the problem–solving process would be observed. Positive emotions tend to be frowned upon or viewed as "childish." Few people exhort optimism like Ray Bradbury: "We are matter and force turning into imagination and will! I am the center of a miracle! Out of the things I am crazy about I've made a life. ... Be proud of what you're in love with. Be proud of what you're passionate about!" (Bradbury, 1982). People who exhibit positive emotions often are regarded as playing, as not being serious.

Yet, playing with ideas is inherent in mathematics problem–solving. Then what emotions should we expect to feel? Lazarus (1986) answers this by saying that the essence of play is that it is highly stimulating.

It is accompanied by pleasurable emotions such as joy, a sense of thrill, curiosity, surprise, wonder — emotions exploratory in nature. Stephen Brown (1971, 1982) is one mathematics educator who has explored ways of promoting those emotions in teaching.

As educators we are concerned about the classroom environment. Are there optimum conditions for encouraging problem–solving activity? "Exploratory activity occurs more readily in a biologically sated, comfortable and secure animal than in one greatly aroused by a homeostatic crisis," observed Lazarus, "The human infant will not venture far from a parent unless it is feeling secure, at which point it will play and explore, venturing farther and farther away but returning speedily if threatened or called by the mother." The novices in our study felt increasingly secure due to the support of their partners, and reported more out–of–class mathematical discussion.

The identification of emotion was through self–report and peer-observation, with no external verification of the visceral component (e.g., measurement of change of temperature or blood pressure, for example) or measurement of intensity or duration. Observers were, however, asked to record sighs, laughter, twitches, wiggles, tapping, yawns, or flushes. A heightened emotional intensity appeared to occur at four stages of the problem–solving process: (1) at the initial reading of the problem, (2) when embarking on the problem–solving process, (3) during immersion in the problem, and (4) when a solution was found or when the session was about to end.

Additionally, there appeared to be a rhythm or cycle during the engagement with the problem: first, attention to the problem or "immersion," then outward attention or "distraction," followed by total immersion, then distraction, etc. Integrated with, and slowing down, the cycle were moments when the attention remained on the problem but the accompanying actions were mere "holding patterns," such as rewriting the question, with almost no cognitive load. These appeared to be times of incubation, cognitive "breathers," rather than times of woolgathering. No stress was involved, and all attention remained on the problem. The next section of this chapter outlines the research procedure for the experts and is followed by a discussion of those results.

EMOTION IN EXPERTS' PROBLEM–SOLVING

The data discussed in this section were obtained over a three–year period in the United States and Canada. In the United States, 12 subjects were professional mathematics educators or were employed in business or technical fields in which they were accustomed to using mathematics. Another six subjects were graduate students in a mathematics problem–solving course at a state university. In Canada, 19 subjects were mathematics educators participating in study groups on the role of feelings in learning mathematics held during two annual meetings of the Canadian Mathematics Education Study Group (CMESG). Except for the graduate students, the participants were all seasoned mathematicians age 40 or older, eight of whom were women. Data were not gathered on the mathematicians' specialty, e.g., topology, algebra.

PROCEDURE FOR EXPERT'S PROBLEM SOLVING

The main activity was paired–problem solving, where one member of the pair worked on a mathematical problem under the observation of the partner. We intended to make as transparent as possible the emotions and cognitive activity during the process. The format of each occasion was the same. Six to 10 subjects gathered together with two co–leaders (John Poland, Carleton University and Rosamond at CMESG; Maria Arrigo, San Diego State University and Rosamond in San Diego), engaged in a general discussion of emotion and then read a list of words and phrases that might indicate various emotions. Our emotions seemed to have more shades of meaning than we had names for. We discussed physical manifestations of feelings, such as laughter that signifies nervousness or yawning that signifies confusion. Subjects suggested that some visceral sensations, such as butterflies in the stomach, could be associated with either positive anticipation or dread.

The participants then paired–up, one person agreeing to be the problem solver while the other was the observer. The problem solvers then browsed through books and handouts supplied by the leaders to choose a problem. Problems were nontrivial but within the knowledge range of the participants. They were chosen from puzzle books by Martin Gardner (1967, 1979) and Mott–Smith (1954) and from books by Honsberger (1970, 1973).

The problem solvers were to work as transparently as possible and to provide a running commentary on their thoughts and emotions. The observers were to pay attention and take notes, to prompt the subject to verbalize whenever there were long periods of silence, and otherwise to remain quiet, withholding information or advice.

Following the previously agreed–upon solution time (15 minutes in the first three sessions, 30 minutes in the last two), all participants assembled to report orally. We sat in a circle, and each observer reported on the emotions and actions witnessed during that observer's partner's problem–solving. Directly after each observer spoke, each solver could correct or add to the report.

The roles were then exchanged: observer became solver, and solver became observer. The new solver chose a problem, and the solving, observation, and reporting process was repeated. The data used in analysis consisted of the written notes from each observer, the problem solver's written work, written observations from the co–leaders, and the author's notes covering the whole–group discussions, observations, and discussions with the co–leaders.

RESULTS OF THE EXPERT PROBLEM SOLVERS

This section describes the emotional reactions that appeared to influence cognitive decisions during four particular stages in the problem–solving process.

1. The Initial Reading of the Problem

2. Embarking on the Problem-Solving

3. Immersion in the Problem

4. Solution to the Problem or near end of Session.

THE INITIAL READING OF THE PROBLEM

In three sessions, the expert problem-solvers were invited to choose their own problem from the books listed above. Their choice had to be made within a few minutes. As a result, problems were chosen that had at least some initial appeal to the solver. In the other sessions, problems were handed to the solvers. At no time did a solver refuse the problem, and solvers seemed to accept their problems with curiosity and positive anticipation. The initial reading of the problem provoked an immediate reaction based on identifying it by type followed by a sense of its difficulty. Two statements illustrate the initial evaluation of type and results. "I anticipate I will enjoy this problem but may not make much progress," and from another solver, "I loathe this type of problem. It is do–able but will require a big effort. I think I will have to go through many tedious decompositions."

The word "do–able" was used often and meant either that the problem was solvable or that progress could be made in understanding the question. As indicated in the second quote, the anticipation of being able to come to a solution of the problem and the enjoyment of working on the problem were not directly linked. All solvers were more encouraged by harder problems than by ones marked "obvious" or perceived as easy. Problems that were perceived as difficult were accorded greater value in the sense that intellectual effort must be expended to understand the problem. The intellectual effort was linked to the emotion of pleasure. Those who felt the problem worth working felt an immediate joy even before proceeding. This joy was a signal to bring all mental force to bear on the problem, which in itself produced pleasure and therefore motivation to continue.

EARLY IN THE PROBLEM–SOLVING PROCESS

The following episodes clearly show that emotions have a functional role in guiding the choice of method. After reading the problem, all solvers began to develop a notation, to draw a diagram, or to write out a hypothesis. They were totally focused on the problem at this stage. This was the beginning of a cycle of immersion in the problem discussed below. At this stage, considerable emotion accompanied the choice of method. In particular, solvers clearly expressed an emotional repugnance for any method which they considered "cheating" or "bad sport." Three definitions of "cheating" emerged. First, cheating, for these experts, meant any method or technique that was perceived as more powerful than the level of difficulty warranted by the problem. Each solver had already made a judgment about the level of difficulty of the problem based on the solver's perception of the type of problem, on how successful the solver had been with those types in the past, and on how much the solver enjoyed that type. For example, a university professor, Solver X, speaking aloud, said "Can I use fancy stuff? … Then I'll use the Jordan Curve Theorem…." Then he laughed in an embarrassed manner, backtracked, and began to reread his notes, saying, "Maybe there is an easier way." Another professor, Solver Y, with a problem entitled, "An Obvious Maximization," spent several minutes resisting using calculus before finally making a grudging commitment to calculus techniques.

Using brute force was considered as bad as using a too powerful method. One solver expressed the feelings of several when, while tapping his pencil in an agitated manner, almost as if he were drumming, he muttered, "I'm annoyed because I cannot see any other way than brute force, and that would not yield for me any understanding of the problem ... there must be an easier way." Consideration of a too powerful method, brute force, or an "obvious method," brought forth comments and behaviors indicative of embarrassment or annoyance.

A less conscious resistance to cheating was seen when solvers imposed ridiculous restrictions on themselves. One solver, for example, had Honsberger's book in hand and was attempting to solve a problem that began, "Use the Method of Reflection to ..." The solver's reaction was, "I understand the problem but I don't know this method ... I wish I could read the chapter that discusses the Method of Reflection." The subject had self–imposed the restriction against reading the chapter. He considered it a form of cheating. Instead of simply turning to and reading the chapter, the subject tried to invent on the spot a plausible "Method of Reflection."

Another solver spent long moments in seemingly aimless thought, finally saying, "I'm feeling a little out of control of the problem ... There are lots of parameters ... There seems to be a lot of ways to define this problem ... I'd like to clarify the problem by asking whoever wrote it." The subject was holding back from defining the problem himself. Finally, with a forced air, the subject said, "I could break it up into cases myself and come to grips on my own terms and get partial solutions ... I've got control back."

In general, self–imposed restrictions or feelings of cheating would slow a solver down until he felt uncomfortable enough to say, "I'm wasting time. I really haven't done anything." Then there would be a squaring of the shoulders and a businesslike assertion to "... take a stand and try to prove it ...," even though this might mean grinding out a meaningless, albeit correct, solution. These excerpts from solvers' comments demonstrate the appreciation of an elegant, efficient, or otherwise aesthetically pleasing method of solution. While the desire to attain such a solution at times seemed quite conscious, subjects seemed less aware of the restrictions they imposed on themselves as a result of the desire.

IMMERSION IN THE PROBLEM

Once solvers made a commitment to a method of approach and made some progress, they became totally immersed in the problem. Their engagement was so complete that it appeared to make them almost oblivious to the observer, the environment, or themselves. At moments like this the solver would be writing on the paper or would appear lost in thought. Any verbalization to the observer was parsimonious. This level of intensity was not maintained constantly, but it was one mode of a cycle of attention to the problem (immersion), followed by attention outward to the observer or environment (distraction), followed by a diving back into the problem.

The distraction mode was prompted by several factors. When the solver paused overlong in appreciation of some success, then the solver's attention would gradually be drawn away from the problem to the environment or towards the solver's own self e.g., "I'm warm in here," or "It's warm in here." Likewise, the jolt of finding a counterexample to a hoped–for truth would prompt the solver to notice the ticking of the clock or the temperature of the room. To sit too long without progress would cause the solver to recount a memory of a similar past experience. Typical solvers, for example, after a lengthy period of frustration, would tell the observer that they were not really very good at this type of problem. Attention was diverted from a focus on the problem, to a focus on the self, sometimes accompanied by embarrassment. This outward attention was brief, usually less than a minute. Solvers would look around, stroke the pen, sigh, scratch, talk a little and then go back into the problem.

Although going too long without progress or action tended to bump the solver out of concentration on the problem, there were some activities that seemed to allow the solver to remain immersed in the problem even though no progress was being made. These activities could be called "breathers"; they were mathematical routines that required little cognitive energy. A typical routine was simply to rewrite the definition of the variable. One solver began, "There are two cases: (a) The problem is solvable, and (b) The problem is not solvable." Almost all solvers who began a problem using x's and y's, came to a point where they switched them to a's and b's, and vice versa. Another routine was to decide abruptly to use induction, and then write out the complete induction hypothesis. The problem–solving process took on the form of the following cycle.

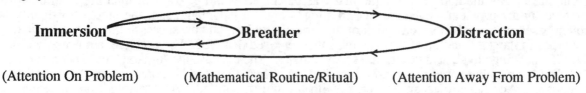

Immersion	Breather	Distraction
(Attention On Problem)	(Mathematical Routine/Ritual)	(Attention Away From Problem)

The mathematical routine of switching variables or writing definitions appeared to provide moments of intellectual relaxation during which the solver continued to maintain an intense state of concentration on the problem.

EMOTION AT SOLUTION OR WHEN TIME WAS CALLED

Most of the solvers were in the immersion mode of concentration on the problem when it was time for the session to end. The emotion expressed by these people was irritation at being interrupted. Almost all of these solvers mumbled, "I'll continue later." They would have preferred to remain solving the problem. Those subjects who were in the distraction mode just prior to time being called, were usually aware that the time was just about up. They tended to sit back, observe others, and wait out the time. They did not work on the problem further, but mentioned that they would return to it at a later time. What became clear was that the attraction to the problem was balanced by the reluctance to allow oneself to get lost in a train of thought only to be yanked out of it.

The overall realization of the time constraint inhibited some problem solvers. An example of this was found in one solver's report, "I feel hemmed in. I do my best by playing around ... ordinarily I would draw pictures and really understand ... build up a pattern." He felt pressure to categorize a solution method quickly. "Without a time constraint I probably would have been more impulsive. I would have guessed and then worked backward. I felt forced to be more systematic, meticulous, and more step–by–step and mechanical. I think I could have solved this in a shorter amount of time if there had been no time limit." Although this solver produced a successful solution, the fact that there was a time–limit produced emotions that inhibited him from becoming as immersed in the problem as he would have liked. His attention was divided between the math–world and present time. Although he thought he could solve the problem in a shorter amount of time, he did not attempt to explore a shorter method. When the timing in itself counts, it is as though the meaning of the problem diminished.

Solvers expressed disappointment if the solution came so easily that little emotion was invested in the problem. Then they would remark, "The problem must have been too easy. I got it. So what's the big deal? I feel let down because I didn't spend a lot of emotion." When the complexity of a problem came like a revelation to one solver, he responded with a BIG smile.

One solver exhibited obvious arousal with eyes wide open, clear face and a slight laugh: "Hey, there's an infinite process ...." Exploration didn't bear out the infinite process, with the result of a clear drop of interest and a rather emotionless settling again into the problem.

EMOTIONAL RESPONSE TO USEFULNESS OF MATHEMATICS BY EXPERTS

A feeling of usefulness of mathematics played an especially important role for the women. Five of the eight women and only one of the 23 men brought up the issue of usefulness of mathematics. Subsequent to this study, one of the women has left her teaching career of twenty years to pursue a degree in another field in which she feels she will be able to make more of a contribution in the world. A different participant clearly describes the strength and influence of feelings of usefulness. This participant was a medical doctor with a substantial background in mathematics. She was able to solve the assigned problem in a short time and with no intense engagement. She was disappointed and felt let down. She insisted her feelings would have been the same had the problem been more challenging. In a loud and agitated manner she described her feelings:

> What would have been a meaningful problem? How come I'm not satisfied? I had
> an expectation about solving that problem that did not get fulfilled. It didn't make
> me happy. There were some moments of tension and some of excitement, but not
> intense. It was entertaining like a Grade C movie.

She had related the value of the problem to the intensity of emotion involved in its solution. She went on to describe the usefulness of mathematics:

> Math has no social relevance to me. I am willing to solve math problems, even
> ready. But it feels completely disjoint from what interests me. I still love it, but its
> importance seems minuscule compared to world problems ... beautiful but frivolous
> to use my mind in this way.

She said that, while it might be useful for other people to do mathematics, there were more pressing issues for her to engage in. Mathematics has an esthetic appeal for this solver. She sees mathematics as beautiful and is attracted to mathematics. She is confident. She was successful in reaching a solution and feels ready and willing to solve others. It seems apparent that the intensity of her medical problem–solving and its immediate applicability have set a threshold of "intellectual utility" that recreational mathematics cannot reach. However, this episode and the comments of the other women show that the emotion attached to usefulness of mathematics can be overriding and can serve as a powerful influence on whether or not to engage in mathematical activity at all. This is in accord with research by Fennema (1989) and others who have shown perceived usefulness of mathematics to be one of the most important variables in gender-difference research.

SUMMARY OF THE EXPERT GROUPS

Emotions influenced the problem–solving process both positively and negatively. Excitement, hope, eagerness, and the "joy of battle" were positive emotions mentioned by the solvers. The positive emotion of hope provided motivation to keep going. Occasionally during problem–solving, the solver lost control of the problem. One solver said, "This is too complicated, too many angles to label," and another, "I feel this is getting a little out of hand. This one and that one cancel out and I haven't used the fact that it's a prime." The solvers had confidence in their own abilities and experience with doing mathematics, so they did not lose hope and did not stop working. Hope, the belief that there is even a slim chance things will work out, helps one continue. Ambiguity nourishes hope. One cannot be hopeful when the outcome is certain.

Solvers who perceived their problem as too easy felt disappointment even before they began to work on the problem. The emotions of those who perceived the problem as a challenge are summarized by one subject as "the joy of mental engagement and the bringing of all mental force to bear in a cohesive way." The mental engagement took on the form of total immersion in the problem. There was a Lorelei seductiveness about it, a delicious slipping off into another world. During discussions after the sessions, people revealed times when they used mathematics as an escape from daily life, such as to ignore the pain of an illness or to avoid thinking about someone. Mathematics can help with depression, as the famous mathematician Kovalevskaya wrote, "I am too depressed … in such moments, mathematics comes in handy, and one enjoys the existence of a world completely outside of oneself" (Koblitz, 1983).

Mingled with the charm of seduction, the math world takes on a dangerous quality, a frightening isolation if one stays immersed too long. Rosamond (1982) has given examples in which problem-solvers felt consumed by a too dominating mathematics. As one distraught mathematics graduate student said with tears in his eyes, "What do you do if you are 80% — 90% mathematics? If you've let yourself become consumed by mathematics so that that is what you are and then you want to let someone get to know you? What do you do when you can't explain that much of yourself to them?" Working with an observer was a protection against becoming dangerously immersed or overly isolated. We wonder how professional mathematicians relate to these faces of mathematical life.

EMOTION IN NOVICE PROBLEM–SOLVING

The second study involved mathematical novices, consisting of six students in an intermediate algebra course, 14 computer science and two mathematics majors in a college algebra and trigonometry course, and 12 mathematics majors in two history of mathematics courses, all at a private university. The study consisted of six activities. In all the activities the students chose their own partners and worked in pairs. The two initial exercises were short discussions of only two minutes' duration. The second two were three–minute visualization exercises. The last two were problem–solving sessions of about ten minutes each. During each activity, one partner was a silent, attentive observer who took notes on the partner. Then the entire group came together and each observer reported. Each problem-solver could amend or add to the report. The observer/solver then switched roles in each pair, and, after the second person worked on a problem, the entire group again gathered together. The data for the research include the notes taken by each observer, the work done by each solver, and notes and tape recordings of the group sessions made by the researcher.

Whereas the format of the exercise was similar to that of the experts, the experts gathered together for the fun of the investigation and activity, while the novices were students compelled to do whatever the teacher suggested.

Having the students discuss the exercise together as a group was an especially useful format. Students would add to and build on other students' comments. A comment by one student would remind another of some idea. This is similar to the experience Casserly reported when she interviewed students in groups while studying feelings of gifted students enrolled in advanced placement programs (1979).

RESULTS FROM THE NOVICES

The following data was compiled from the observation reports of the paired problem–solvers, observations of the teacher, and audio tapes and notes from the discussions following the problem–solving sessions. While conclusions have been drawn, these must be regarded as tentative, intended to be provocative of further research. The results will be described briefly and compared with the data from the expert solvers.

INITIAL EXERCISE

The initial discussion activity was chosen with two purposes: to provide students with practice in paying attention to a peer, an important experience for an observer, and to provide practice in describing feelings about mathematics to a peer, an important experience for a problem-solver in this research. The initial activity, repeated on two occasions, was for student pairs to spend two minutes each, one person talking and the other giving attention, discussing a statement provided by the teacher. The first statement was, "Why I like mathematics," and the second statement was, "Why I am an excellent problem solver." In the general discussion that followed, several students agreed with the algebra student who said, "At first I said that I wasn't a very good problem solver. But then, I still had more time and so I began to think of ways in which I was." An all A–student who is very good in mathematics said, "I felt like I was my own cheerleader. It felt good."

In the brief paragraphs students wrote about the activity, several students expressed a sense of relaxation. For example, "The short session helped to relax me a little. It was enjoyable to talk for a few minutes and relax. It kind of helped my mind to relax, and then later some items that I found hard to understand kind of came together all by themselves."

Another student wrote, "I feel it (the activity) creates a positive atmosphere and provides a short escape while not totally losing focus on mathematics. I liked it."

Most students found the activity unique in that this was the first time they had ever thought about their feelings with respect to mathematics. An algebra student wrote, for example, "I felt kind of funny doing tonight's exercise, but its a good idea to bring out the positive points of math and it surprised me how many things I like about math."

Another student made a similar statement, "I feel that it released feelings I have about mathematics by saying them out loud for the first time. Also, by listening to another person's thoughts on why they find mathematics interesting. At first I felt stupid, but now I can see … some more potential in doing this …."

A computer science student wrote, "I had never verbally given my reasons and feelings about mathematics to anyone. This is important to get a real, solid reason for being here, because other curriculums would be less demanding. I know from experience that you must set out your goals to develop commitment — and this exercise starts me along the process of developing strong commitment." Another student seemed to second this opinion by writing, "I believe this is a good idea because if a person does not feel good about himself then he is not going anywhere no matter how much he learns."

A mathematics major summed up the general positive reaction to the activity by writing, "This (activity) is a great idea. We are not always willing to open up to each other but … why not? Working on feeling good about yourself can never be bad!!"

Some students were more enthusiastic than others, but every student was intrigued and willing to do the activity, including the few for whom English is not their first language and who expressed some difficulty with communication. The activities provided a relaxed way to begin to discuss feelings. It was deliberately kept short so as not to intrude on other classroom work.

VISUALIZATION AND TEXTBOOK PROBLEMS

The last sessions were preceded by a short, teacher–led discussion of feelings and mathematics. The teacher pointed out that feelings are difficult to describe and that sometimes it can seem as if there are not enough adequate words. She asked the class to provide words and phrases that expressed their feelings about mathematics. The teacher listed these on the board as quickly as the students called them out. The list of words and phrases can be found in the appendix.

In the third and fourth sessions, a problem was provided by the teacher with the instructions that no pencil-work was allowed for at least three minutes. That is, the solver was instructed to visualize the problem and perform stretches, rotations, and other transformations mentally. Students relaxed with eyes closed while the teacher read the problem aloud. They were invited to ask questions for clarification. One person in each pair described thoughts and feelings to a partner. The two visualization problems are described in the appendix. Students tended to find this activity frustrating, and it did not appear to enhance the subsequent pencil-and-paper solutions.

For the last two sessions, each solver chose either a homework–type problem from a textbook or a problem from a puzzle-book. The time allocated for working on the problem was about ten minutes, followed by the group's gathering to discuss the observations. Then the roles of solver and observer were exchanged and the process repeated. The discussion of the novices will largely parallel that given above for the experts.

COMPARISON OF NOVICES AND EXPERTS

Some of the emotional aspects of problem–solving exhibited by the experts were observed in the novices; however, it was clear that the novices had much less mathematical experience on which to draw, and more recent negative experiences.

THE INITIAL READING OF THE PROBLEM

The students in the college algebra and trigonometry course were given their choice of three problems from their textbook to solve. Reasons for choosing a particular problem varied, from comments such as, "It seemed like something we had been working on, familiar," and "It makes sense and I can draw the starting picture," or "It looks doable," to "Basically, because it is challenging." One student chose a clock problem because she did not want an airplane problem; another liked aviation. These students had some sense of the type of problem they chose, and they tried to judge the amount of effort that would be required. For most students, the problems turned out to be more difficult than they expected. Initial confidence was replaced by frustration at being stuck, anxiety over not being the first finished, confusion because an apparent answer didn't look right, relief upon realizing some algebraic miscalculation or relationship error, and a sense of satisfaction or letdown when time was called.

The variety of feelings as described in these comments written by one observer are typical.

O (Observer): Why did you pick this problem?

S (Solver): It was most recently covered in class.

O: How do you feel about math?

S: I have no strong feelings about math.

He reads the problem.

S: Oh, man! I'm no good at these!

He draws a picture for the problem, then says, "That's easy." He explains the variables, puts them into the Law of Sines formula, and starts to solve.

S: I feel very happy now, I've got it! … I'll check.

He rereads the problem, then loses confidence in his answer.

S: I'm not sure if I solved for the right thing.

He looks, re–reads, re–draws, tries many re–draws, says, "Oh, I've got it!" and attacks his calculator with gusto. He works out his new idea and solves the problem.

S: I feel pretty good about this answer.

He says, "It is very satisfying to solve a math problem." and gives an arm pump.

On the same paper the problem solver has added, "I didn't pay too much attention to the observer. It was rather motivating for me to be observed."

One of the observers of a mathematics major reported that her partner realized the problem was going to be tedious. "Not that he couldn't find an answer, but just that it would take a lot of time and a lot of brain power. He seemed impatient, as though, 'I don't want to have to sit here and have to do this and take all the time to do it.'"

As with the experts, the novices had some evaluation of the type of mathematics problem they had chosen. After reading the problem, they anticipated pleasure or frustration, and they had a sense of the effort needed for results. Whereas the students' limited experience meant they had fewer methods of solution available to them, they still made an emotional appeal to a sense of appropriateness.

The mathematics majors were offered puzzle-books and textbooks from which to solve problems. Ten of the twelve chose a problem from a puzzle-book. One student explained, "I like word problems … and it's a real world type and I like that sort of thing." She began the problem and then said, "I changed variables. I changed the way I put my equations. I thought I'd try simultaneous equations. Then, I said let's try some numbers. The minimum it can be is this …."

T (Teacher): When you switched variables, did sometimes you use a's and b's and other times x's and y's? Did you switch?

S (Solver): Yes.

Almost every problem-solver reported changing variables or re–drawing a picture during the solving episode.

> T: Sometimes people don't want to use a more high powered method than the problem deserves. Is that what was going on with you when you said first you were going to try simultaneous equations and then you seemed to back off and decide to just try some numbers?

> S: Yes, it shouldn't be that difficult. It (the problem) sounded easier than this so I shouldn't have to use something like that.

In one case an almost uncontrollable negative reaction during the initial reading of her problem severely inhibited an algebra student. She reread the problem repeatedly both silently and aloud, but did not appear to be thinking about the words. She admitted being anxious. "When I look at a math problem that seems really different from one I've just studied in the book, I get so anxious. I want to solve the problem but I don't know how to start. I feel bad and then I feel bad about feeling bad." This student spent much of her time telling the observer about how she could understand a problem when it was explained in class, but how she would struggle unsuccessfully alone at home. Her anxiety reaction is not an uncommon one. Turner (1988) describes the rehearsed pattern of failure that becomes "not only well established and familiar, but an acceptable, and ultimately relied upon, event." Just as the confidence expressed by the experts provided them with a positive anticipation toward reading the problem, the lack of confidence expressed by the algebra student blocked her from even trying to make sense of the problem. She is reacting as Mandler (1989, p. 244) has suggested, with autonomic arousal affecting her access and retrieval of knowledge stored in memory. "The occurrence of strong negative affect produces an immediate attempt to remove the reason (the cause)." The caring attitude of the observer made a difference for this student. (All partners stayed together for all activities.) Her partner gently reminded her that she had excellent problem–solving skills, and he knew this because she had told him. After this small positive affirmation from her partner, the student read the problem through once more, very slowly, drew an accurate picture, and mumbled as she worked, "Yes, I have excellent problem–solving skills." She determined that the triangle she had drawn was too large but was unable to continue because the time was called. What is particularly interesting is how quickly the intervention of the observer influenced the problem–solver's ability to make progress on the problem. The role of the observer will be discussed further below.

IMMERSION IN THE PROBLEM

Each of the mathematics majors reported being focused on the problem to the exclusion of anything else going on around them. They described the modes of immersion and distraction. Student B reporting on Student T says, "He did not want to stop when time was called. Basically, he was so intrigued with the problem that I was working on that he just went right back into it … he was enthusiastic and using new variables and talking through the whole thing … he was concentrating and very methodical and checking things off, working and checking, and he kept this up for some time, and then paused, rubbed his chin and said, 'I'm tired of this. I'm going to try something different.'"

"He was looking for new relationships and the time went by very quickly. I don't think he noticed that anything else was going on in the room. And I felt the same way working on the problem. You have no concept of anything else going on. It is real focused. Anyway, he was determined to keep going, he didn't want to quit (after time was called). He still had a lot of energy left."

> Student T: What is it that bumps you out of that concentration state?

> Student B: For me it might be someone's real loud voice. It's something startling that will get me out of it …

> Student T: What is it that gets you out of it when nothing startling is going on and you are doing the mathematics? What's going on in your problem–solving that gets you out of it?

> Student B: Solving the problem or getting to a point where I really don't know where to go next, and then I just have to get up and walk away. But we didn't have enough time for me to get to that point yet because there was still energy left for this problem.

> Student T: Yes, I agree. Either solving the problem or you are just going in circles and so your mind wanders, you are thinking of different ways to approach the problem and then your mind wanders and you think of the beach … Your mind

wanders in search of a new idea and it happens to take your attention away from the problem in the course of wandering.

Negative self-talk also can take attention from problem–solving, as reported by Student W, a mathematics major.

> Student W: If I'm concentrating on a problem and I get to a point where I'm so frustrated, then the thoughts begin to run through my mind: Look at this, you can't do it. How stupid. It's ridiculous that you cannot do this problem. Why can't you remember? Gosh, anyone can do this, you moron. There is a lot of negative talking to myself. That's when my thought is that I just got to give up. It's just a feeling. It's not giving up forever. It's just a feeling of giving up until I can regroup and come back to it. But for now, forget it.

> Student T: How long does this last? For minutes or for days?

> Student W: Ahhh, just at the point that I feel it. Once I put the pencil down and go do something else, I don't feel it anymore. I know that when I come back to it, if it's something I've done before then I'll remember it. If it's something I've never done before then, then I may not get it at all.

There was more dialogue between the novice solvers and observers during the problem–solving sessions than there had been with the experts, and observers had more difficulty restraining themselves from giving hints or explaining what they would do. One solver reported being angry that his partner wouldn't help him. He was convinced that together they could have solved the problem. Observers would remind the solver of a problem–solving heuristic now and again (some of the classes had compiled a list from those of Pólya, Brown, Schoenfeld, and others) or ask leading questions.

SOLUTION OF THE PROBLEM BY NOVICES OR END OF SESSION

Most students expressed irritation when time was called and said (as did the experts) that they would continue later. Student J had solved one problem and was in the process of working on a second.

> Student J: I was still coming up with a way to solve as opposed to an actual solution … There was some impatience because the method wasn't jumping out at me, but beyond that it wasn't a real emotional experience.

> Teacher: Was the first one, where you solved it, a bland experience?

> Student J: Actually, when I read the problem, to be honest, I felt pleasure because it was so obviously clear that it was solvable quickly. There were not many variables. It just wasn't that complicated. Although I didn't know what the immediate answer was, I felt like I could solve this one right away and there was pleasure in knowing that. Then I went ahead and solved it and I got confirmation of that because I did solve it.

> Student M: When I finished mine I thought, "Oh, I can't believe I did it. It can't be right." I looked at the answer thinking, It can't be that easy.

> Teacher: When you finished that problem, did you have a sense of accomplishment, puzzlement, or what?

> Student M: Puzzlement — accomplishment too, but "Did I do it right?" I figured out the problem according to the parameters I made, but I didn't know if they were what was wanted.

> Student R: Some people have said they feel a sense of let–down–ness if they solve a problem that is too easy. Does that make sense to anybody?

> Student W: If it doesn't seem easy at first, if when I first read it I say, "Hmmm, am I going to be able to solve this?", then if I get the right answer, but if it takes me only five minutes to do, then that feels good. But if when I read it it is just a word problem, I can do addition of apples and oranges, then that's not hard at all, then that's not …

>Student B: If there's not a little bit of challenge, then there's ... not a point of doing it ... the challenge of overcoming that ...

>Student M: For me, there's a big let down. You get into the problem, getting all ready, your brain is mustering its all to throw at this problem ... Wait a minute! I could have done it at half–strength! I didn't need everything on this one.

As with the experts, the emotional charge for the novices appeared in direct proportion to the perceived difficulty, except when the problem was seen to afford a quick, simple, solution.

EFFECTS OF THE RESEARCH PROJECT ON SUBSEQUENT STUDENT BEHAVIOR

Conversations with students after the course showed that the research project had introduced students to a new way of working problems which appeared to have a substantial effect. It was clear in the class that the pairs began naturally to turn to each other to work problems together. They also talked to each other about their math outside of class. Students reported that, when they were working alone on a problem, they could imagine their partner with them and what the partner would be saying. When I (the author) asked what this would be, the response was that the partner would be supportive with statements such as, "Come on, you can do it," and inquisitive with questions such as, "Where are you stuck? What part of the problem are you on? Convince me you are right." Students said that this calmed them down and helped them back away from their confusion and look at the problem afresh. (In class, we had spent some time working problems and identifying the "Phases of Work" described by Mason, Burton, and Stacey [1982, 1984], and using the "What if Not?" techniques of Brown (1982).) These responses confirm the results of Clement and Konold [1989].) who found that "having students work in pairs alternating between the roles of solver and listener helped students internalize the roles of solver/listener and learn to carry on an internal and critical dialogue with themselves." Students in this study reported that they felt a heightened responsibility to be able to explain, and verify their solutions.

In summary, paired problem–solving begun in the classroom was continued by the students outside of class. Students wanted to come to class in order to see their partner or to support their partner. Students began to internalize the roles of solver/observer, to "hear" the positive affirmations of their partner, and to ask themselves the questions they knew their partner would ask.

EMOTIONS, BELIEFS, AND THE ROLE OF THE CLASSROOM TEACHER

Mathematics can be considered as a way of viewing the world, and its meaning and process are subjective and individual. On the other hand, mathematics is a common language with rules and precision. From Cobb (1989) we have, "The complementarity that seems endemic to mathematics education theorizing expresses the apparent paradox between mathematics as a personal, subjective construction and as mind–independent, objective truth." Many students tend to see mathematics as precise, rule–driven, objective truth rather than constructible knowledge. The terror that comes over some students when faced with a mathematics problem can be related to their belief that the problem is an all–or–nothing situation, "I either know IT, or I don't, and there is no use in my trying to figure IT out" (Copes, 1982, p. 24). In addition, students at early stages of intellectual development with respect to the discipline will ascribe to the teacher the sole Authority and repository of the Truth (Copes, 1991; Perry, 1970).

Furthermore, whereas the goal is to assist the students in the construction of their own mathematical knowledge, teachers still tend to feel fully responsible for the mathematical learning of their students. Fischer (1988) describes mathematics teachers as tending to "identify themselves with the subject matter and present mathematics as if they were totally convinced of its truth and worth." Thus, there is a sort of collusion going on in classrooms between students who view the teacher as the Authority and sole distributor of mathematical Truth, and the teacher, who has assumed the embodiment of mathematics.

Paired problem–solving can help disengage the collusion. Research by Perry, Copes, and others on the Perry scheme of cognitive and ethical development and on Kohlberg's model of moral development shows that students did not move forward in development when they were in an environment more than one stage later than their own, but did move forward when they were in an environment just one step beyond their current stage. Pairing students puts them with peers that are usually at the same or slightly later stage, and thus can be a force toward cognitive and ethical growth.

Other instructional environments and teaching techniques help remove the burden of full responsibility away from the teacher. Carefully designed computer labs such as the one designed by Dubinsky (1988) and Schwingendorf at Purdue propel the students into relying more heavily on themselves, the computers, and their lab–mates. Teaching the students to read mathematical textbooks and literature, as Clarence Stephens has done at Potsdam, provides the students with an "Authority" other than the teacher.

Along these lines a wide variety of successful programs was described at the first National Conference on Women in Mathematics and the Sciences (Keith and Phillip, 1989).

PHENOMENA COMMON TO BOTH EXPERTS AND NOVICES

EMOTIONS AND THE INFLUENCE OF THE OBSERVER

Contrary to almost everyone's expectation, being observed while working on the mathematics problem was a positive experience. Both experts and novices reported that being observed, "Made me feel important" or "... that what I was doing was important." An expert reported, "I felt honored that another person was taking the time to observe me." Another expert said that the analytic intimacy she felt while being observed was one of the most poignant mathematical experiences she has ever had. "It felt intimate to have someone committed to watch the inner workings of my mind." Even those who felt frustration with the mathematics problem enjoyed the process of being observed.

An expert related the process to the testing situation. "A test is an almost random set of narrow problems where one thing must trigger another. It is not about figuring things out. Test questions do not show that math is a process." This expert did not solve the problem, but was not intimidated by being observed. "The observer could hear that I have math training. He could see how my math mind works, how I assimilate information, manipulate, and use an arsenal of strategies. This is so much different from taking a math test where I am not tested on how my mind works. On a math test, I could expect not to be able to show what I know. I would feel shame."

One of the mathematics majors commented, "To be perfectly honest, there is a thrill to the risk because if you are unsuccessful you are damned . So there is a certain amount of thrill that goes with the risk and so it becomes very positive when it reaches a point in problem–solving when I can see how to solve the problem and I can say, 'Oh, accolades are coming.' That part of the thrill becomes positive. And, of course, there is the other side, and I've experienced it often, and that is, 'Oh, I'm not going to get it in time!!' And know I have the potential of being sent to the corner over there ... but there is that risk and the elation that comes with risk."

Most of the solvers continued working on the problem longer than they would have, had they not been being observed. One novice explained, "I felt I owed it to my observer to do my best." Almost all claimed to like their problem more the longer they worked on it, and those who did not like the problem initially began to like it and to get interested in it. Without an observer, these solvers might have quit.

THE EFFECT OF REPORTING IN A GROUP: EXPERTS AND NOVICES

The group-reporting session provided an effective mitigation of math anxiety for some solvers. An expert, a math professor, revealed, "The most moving part of the exercise was hearing the observer say what I'd done. I did not feel intimidated. I didn't get any of the bad response I expected. The observer demystified my emotional and intellectual engagement by simply listing what I did: 1, 2, 3, 4. This cut it down to size, gave it true proportion." The emotions felt by this solver, her "private agony," were reported in a matter–of–fact tone by the observer. The solver was surprised; and later said her reaction was, "Is that all? Is that all I did? Well, that wasn't so much." Hearing her emotions described flattened out their intensity.

The observers found that the process of observing appeared to diminish some of the secret charge of their own anxieties. "I could recognize my feelings in the other person, and could see how the feelings influenced what he did," reported one observer. "I kept thinking, 'Why is he spending all this energy fussing about it? Why doesn't he just get on with it?'" Observing the emotional energies of another person helped observers to evaluate their own.

A mathematics major provided another positive benefit to the group-reporting process. "When we are working the problems, we are doing stuff that we don't realize we are doing. My observer reported that I was being methodical, and I was. But I didn't really realize I was doing that. And someone else was talking out loud and didn't realize that they were. I don't know if it is beneficial or not, when you don't know and then someone is watching and tells you, then you do know. You see what you are doing that you didn't know you were doing."

Student B: You learn that other people feel the same things as you so it can build a relationship within the class. You also learn how to approach problems, for example, almost everyone in this class read the problem at least four times.

Student M: Other people's emotions are usually a good mirror for oneself. What are they looking like? What kind of body language are they sending?

Student J: One of the things that is coming quite clear is that this whole process really has quite an emotive aspect to it. It's not quite as detached or unattached as I had previously thought it was.

One solver reported that being observed enhanced his precision. He felt he wanted to perform well. Both experts and novices reported an initial reluctance to free-associate ideas in front of an observer who might have already mentally solved the problem, or who might be bored, and some solvers wanted to talk the problem over with the observer, or would look up at the observer hoping for confirmation. All solvers reported an increase in interest in the problem and in solving it due to the presence of the observer and due to the act of reporting.

CONCLUSION

The research appeared to show that emotional elements interact with cognitive processes and exert strong influences during mathematical problem-solving. Solvers anticipated leisure or frustration upon first reading of the problem. They anticipated the amount of effort that progress or results would require. This evaluation and anticipation helped determine the choice of method. Solvers' (especially experts') concern with not using a too-powerful method sometimes caused them to put too many restrictions on themselves. Solvers alternated between immersion and distraction during solving, with emotional satisfaction relating directly to problem difficulty.

The presence of an observer during problem-solving was a highly pleasurable experience that appeared to support persistence, sense of self-worth, and the development of a cognitive and affective monitor.

The process of coming together for the observers to report to the entire group and to hold group discussions helped students see the emotive components of what they were doing. This exercise might be especially useful with future teachers, who may need help in recognizing some of the positive and negative student emotions.

The results may be helpful for helping the teacher design an environment to accommodate student diversity and promote cognitive and ethical growth. Future research may examine the subtle interaction between cognitive processes and specific emotional constructs such as value, motivation, hope, or, as Silver and Metzger (1989) have done, aesthetic values.

APPENDIX I
TWO VISUALIZATION PROBLEMS

1. Imagine a belt around the equator of the earth. The belt follows the contours of the earth and is snug. Now, imagine adding 40 feet to the length of the belt. Evenly distribute this added 40 feet all around the earth. First the belt was snug, now it is loose. What can fit between the belt and the earth?

2. Locate the point A(3, −3) on the coordinate axes. Draw a line through this point, making a triangle with the coordinate axes of area 6 square units.

APPENDIX II
LIST OF FEELING DESCRIPTORS DEVELOPED BY NOVICES

I had introduced this activity by saying that some people liked mathematics because it seemed clean, with only one right answer, unlike poetry or literature, which is open to interpretation. Student B, a mathematics major, said that, while it is true that she used to think that mathematics had all the right answers, and she had confidence she could find them, now, as she has gone on in mathematics, it seems to her to be more and more like poetry, a sense that it is aesthetically pleasing.

Student J described knowing that everything he was doing was dove–tailing to a final answer, even though he didn't know what it would be. He knows and doesn't know simultaneously.

> Student M: When I work hard on a problem, I feel mentally taxed, like I have been exercising my brain. I need a mental break.

Students called out these words or phrases as quickly as I could write them on the board.

aesthetically pleasing
simultaneously knowing and not knowing
mental exercise

satisfaction	concentration	right/wrong
frustration	exaltation	anger
focused	disappointed	confused
excited	industrious	thrilled
accomplishment	hesitation	validation

I mentioned a "lull period" during which a solver waited to find what thoughts would come next. Students identified with this and suggested: incomplete, static, and blocked. One student described it as "Working a process and the process is working and you know what comes next. You are moving from point A to point B." Another student added, "A sense of growth, anticipation, relief." Then I mentioned "isolation" and another said "a feeling that everything is sucked into the problem–solving process."

PART IV
RESPONSES

Baxter offers an examination of students struggling to learn the web of concepts that we usually call "the concept of set." That it is multilayered and more complex in its actual use than communicated by its traditional definition becomes clear in her description of students building and exploring sets in the ISETL medium. Her documentation consists of segments of transcripts of interviews with her students in a teaching context — the interviews are directed towards the promotion of learning, so we see evidence not only that concepts are being formed and elaborated, but we also see the specific problem contexts that give rise to this conceptual growth. In a sense then this is action research, directly embedded in the act of teaching, directly embedded in the context of a college course. In effect it is a careful job of teaching and documentation, with the explicit objective of improving the teaching and learning of that course.

Clearly, it is a beginning, for the process of building general and theoretically coherent explanations of particular phenomena must be rooted in the phenomena themselves. Regularities must be exposed; invariants must be identified. In this paper we see the beginnings of some regularities based in the growth of conceptions of set from a simple unordered collection of items, where the items are conceived as "atoms," to conceptions where the items are more complex and varied, often with their internal structure "exposed," to sets defined by increasingly complicated conditions. The last segment of the paper concerns sets and functions, with additional cognitive complexity associated with thinking of functions as members of sets as well as thinking of them as sets themselves, where normally they are defined in terms of some operation.

As a beginning paper, it precedes the development of an organizing theory. It stays close to the phenomena and, more importantly, is written in the language of mathematics rather than that of cognitive psychology. That is, the phenomena are treated in terms of mathematical content.

Much work needs to be done, work that would build the theory framing the regularities and invariants. Such a theory would also need to help clarify the role of ISETL in the learning process. What, exactly, is the role of an interactive medium, as opposed to an inert one? It is easy to recognize the role of interactivity in promoting certain forms of cognition, in particular the kinds of cognition associated with anticipating a system's output. But, given an interactive medium, which are the particular features of ISETL that serve which particular pedagogical and curricular purposes? Are there particular ways that functions and sets are displayed or evaluated that have particular impacts? Are all of these desirable. How about graphics and arrays, for example? Interesting questions abound regarding the close relation between ISETL and traditional mathematical notation. The fact that mathematical notation puts a premium on abstractness and conciseness pulls its users in a certain direction. Are other relations between the general and the particular possible or desirable? Does ISETL provide alternatives? The fact that mathematical notation evolved in a static, inert medium has led to many of its characteristics, especially its reliance on character strings, lists, et cetera. Dynamic, interactive media offer other representational dimensions, especially time and more directly visual notations, e.g., various forms of graphics. Where is the ISETL experience in this larger representational context?

Understanding How Students Acquire Concepts Underlying Sets

Nancy Hood Baxter
Department of Mathematics and Computer Science
Dickinson College
Carlisle, Pennsylvania 17013

Introduction

A major emphasis in mathematics education is being placed on the development of learner–centered materials. The problem is that most students do not think about mathematical ideas in the same way that mathematicians do. As a result, before materials that support the students learning process can be designed, research should be performed to try to determine how students acquire abstract concepts. These investigations might involve:

- observing students in the classroom setting;

- reviewing the results of examinations, homework, and questionnaires;

- interviewing students;

- making use of previous research results; and

- applying our understanding of the relevant mathematics.

Based on the outcome of this research, supportive instructional activities can then be designed to help students construct mathematical ideas on their own. These materials include:

- paper and pencil exercises;

- discussion activities;

- computer tasks;

- written instructional materials; and

- verbal explanations.

This two–phase approach — research followed by design — is an iterative process. The materials are revised as our understanding of how students learn becomes clearer.

This paper describes the application of this research process in an attempt to gain insight into how students acquire concepts underlying sets. In particular, we interviewed students who had studied sets in a discrete mathematics class, where they had performed specially designed computer activities using the mathematical programming language ISETL (Interactive SET Language). We then analyzed their responses. This paper describes the effectiveness of using ISETL as a tool for teaching and learning, gives examples of ISETL–based activities students had performed prior to the interview sessions, details the interview questions, and summarizes our observations.

What Is ISETL?

ISETL is a powerful, high–level, interactive mathematical programming language whose syntax is very similar to standard mathematical notation. It was developed as a tool for teaching and learning the traditional topics in discrete mathematics: sets, functions, relations, graphs, prepositional and predicate calculus, induction, combinatorics, groups, automata theory (see Baxter et al, 1989; Levin 1990). The language was implemented by Gary Levin based on the compiled language SETL, which was designed at the

Courant Institute under the direction of Jack Schwartz (see Schwartz et al, 1986). ISETL is written in C and runs on most micro–computers and main frames. It is a functional language and supports a variety of constructs, including heterogeneous sets and tuples (sequences) of varying lengths, set and tuple formers, existential and universal quantifiers and functions as algorithms (procedures) and as sets of ordered pairs (smaps). In addition, the language supports functions as first-class objects, the standard control structures, infinite precision arithmetic, some graphics features, and a friendly user interface. Most importantly, however, since ISETL's syntax is similar to standard mathematical notation, students are able to express abstract mathematical concepts in ISETL in a way that feels natural to them. Moreover, writing ISETL code fragments helps students to write precise mathematical statements.

ISETL is interactive. Whenever you give ISETL some input, it does whatever it has been instructed to do, gives the appropriate output (if any), and then indicates that it's ready for more input. One of the things we encourage students to do is to "Think ISETL!" — that is:

- *Enter* an expression or a statement.

- Before pressing the return key, *predict* what ISETL will return (if anything) by thinking about how ISETL might process the information that it has been given.

- Press the return key and *compare* the prediction to the actual output.

As a result, students begin to develop mental images associated with various mathematical processes. Moreover, we find students get very involved trying to correlate their predictions with the results that the computer actually gives, especially if they work together in pairs. In this way they tend to focus their mental energy on appropriate mathematical issues.

WHAT WE HOPED STUDENTS KNEW

Many students can tell you that a set is "a collection of objects." Too often, however, this definition is just a string of words that has no particular meaning. Prior to the interview sessions which we conducted, students had completed computer exercises using ISETL that were designed to help them to:

- think of a set as a collection of objects (unordered, no repetitions) that can be represented by a list;

- think of a set as an object that itself can be manipulated;

- form a mental image associated with the process of constructing a new set from an old one — that is, think about iterating through a domain set, testing whether an item satisfies a given condition, and if so, evaluating a specified expression and placing the result in the set being constructed, and

- form mental images associated with set operations (union, intersection, difference, and tests for membership, equality, inclusion).

To help accomplish these goals, students had been asked to "Think ISETL!" while writing short ISETL code segments and while duplicating terminal sessions such as the one that follows. (Note: The ">" symbol is the ISETL prompt and the "$" denotes a comment. Input to ISETL is in boldface print and output appears in plain type.)

```
> $ Special sets of integers
> {-3..3};   {0,2..11};   {10,8..0};

        {3, 2, -1, 0, -3, -2, 1};
        {0, 2, 4, 6, 8, 10};
        {8, 4, 0, 6, 10, 2};
```

Comments: What's happening here? Notice that it appears that ISETL randomly outputs the elements in a set. Students — and some of our colleagues — always seem surprised by this, but after thinking about it for a while realize that this is the way things should be, since the items in a set are unordered.

```
> $ Representing a set by a list
> S := {"a", 2 + 3, true, {1,2,3}};
> S;

        {5, true, "a", {3, 1, 2}};
```

```
> #S;

        4;
```

Comments: Observing that a set can be an element of another set helps students to start thinking of a set as a mathematical object. This idea is further reinforced by using the built–in cardinality(#) operator to show that the number of items in the set **S** is four (even though many initially guess that the cardinality of **S** is six).

```
> $ Repetitions?
> {1,2,3} = {3,2,1,1,2,3};

        true;
```

Comments: Seeing that the sets **{1,2,3}** and **{3,2,1,1,2,3}** are equal helps students to understand the fact that sets do not have repeated items.

```
> $ Set formation
> P := {x**2 : x in {-9..11} | x mod 3 = 0};

        {0, 9, 36, 81};


> Q := {x**2 : x in {-9..11} | x mod 2 = 0};

        {0, 4, 16, 36, 64, 100};
```

Comments: Encouraging students to think "loop–test–evaluate" when constructing new sets from old ones helps them to form a mental image associated with the process of forming a set. For example, in the case of constructing the set **P**, students are encouraged to think about the variable x looping through the items in the domain set **{-9..11}**, testing whether x is divisible by **3**, and if it is, evaluating the square of the item and adjoining the result to the new set being formed.

```
> $ Set operations
> P inter Q;

        {0, 36};


> {1, {3}} union {"AMS/MAA", 1, 3};

        {1, {3}, "AMS/MAA", 3};
```

Comments: Asking students to think about how ISETL might process the information it has been given helps them develop mental images associated with the various set operations. For instance, some students think about ISETL finding the intersection of sets **P** and **Q** by first finding the items in **P** and **Q** separately and then determining what items are in both. Others, however, observe that **P** and **Q** have the same domain set, and realize that an object is in both **P** and **Q** if and only if it satisfies the condition for **P** and the condition for **Q** — that is, if **(x mod 3 = 0) and (x mod 2 = 0)** or **x mod 6 = 0**.

THE INTERVIEW PROCESS

Our goal was to ask probing questions and try to get inside the students' heads in hopes of answering the following types of questions:

- What cognitive difficulties are they having?

- What in terms of mathematics do they need to know in order to answer a question?

- What is going on in their minds as they think about various concepts?

- What is their understanding as to how ISETL evaluates particular input?

In an effort to answer these questions with respect to students' understanding of the concept of sets, fourteen students who were enrolled an ISETL–based discrete mathematics course at Dickinson College responded to a specially designed list of questions. The students answered the questions on their own with no time limit, and then we talked one–on–one about their responses and tape-recorded the session.

INTERVIEW QUESTIONS. SOME STUDENT RESPONSES AND OBSERVATIONS

Understanding sets: First level

Goal 1.1. Think of a set as a collection of objects in some kind of container – a bunch of things that can be represented by a list.

Questions. Is their schema strong enough to allow anything which they think of as an object to be in a set? Moreover, after having studied sets, is their schema strong enough to include something that they didn't think of as an object when they were first introduced to sets, such as a function?
Note: Students had seen examples of functions, but had not studied them yet.

Problem #1.1. Suppose the following has been given to ISETL as input:

```
F := func (P,Q);
    return P impl Q;
end func;


G := func (P,Q);
    return (P and Q) or P;
end func;
```

(a) What is the meaning of the following two assignment statements?

```
S := {F};
T := {F,G};
```

(b) What will appear on the screen as output if the following input is given to ISETL, where T is defined as in part (a)?

```
for x in T do
    print x(true,true);
    print x(false,true);
end for;
```

(c) After all the above statements have been entered, what will appear on the screen if the following are given to ISETL?

```
random (S)(true,false);
random (T)(false,true);
```

Observations. In their written responses, half the class thought of a function (represented by an ISETL func) as an object and consequently answered parts a–c of problem 1.1 correctly. Members of the other half of the class, as a result of struggling with their responses during the interview sessions, were able to expand their cognitive schema to include a functions as an object. They were then able to use their new understanding to correctly answer parts b and c.

For example, consider the following written responses to part a of problem 1.1, from students who initially did not think of a function as an object:

> [The sets S and T] have no meaning since F and G are expecting values to be passed
> into them.

> S is the set that receives F in it, so S is a set with a Boolean value in it.

S := {P impl Q}

When queried, almost all these students gave the correct verbal description of the answer, "S is the set containing F." During our discussion, each student came to understand that F *itself* was an object in the set — they incorporated the idea of a function as an object into their schema of function. For example, consider the conversation with J:

N. Why don't you read the assignment statement to me?

J. S gets the set containing the function F.

N. Oh, S gets the set containing the function F. Exactly. But that's not what you wrote. You wrote, "S will contain the true or false value returned by F."

J. To actually have something inside the set you have to run the function.

N. You don't think it would be okay to just put the function itself in there?

J. I wouldn't see where there would be any use to that.

N. But that happens to be exactly what happens here.

J. Oh, S contains *the function!*

N. What about the set T?

J. It's the set containing the functions F and G.

Or let's look at B, who struggled a bit more:

B. This is a bad page here. I know what's going on, but I just don't know how to express it.

N. Tell me about the set S.

B. S is going to be the true or false values of F.

N. Read the statement again.

B. S equals the set of F.

N. Ahhh … What's in the set S?

B. Ummm … Boolean expressions … F returns a Boolean expression.

N. You're right, F returns a Boolean value, but what exactly is the object in S?

B. P and Q?

N. I don't see those guys there. Let's start again. How many items are in S?

B. One.

N. What is that one object?

B. (laughs) I'd say true or false, which is one of them.

N. You said before that "S equals the set of F." Could it be that F itself is what's in S?

B. Yeah. But what *is* F?

N. F happens to be the name of a function with a given definition … if F has arguments then it has a value … but what this says is that S just contains the object F.

B. Okay. (pause) What if in ISETL you just typed **F; ?**

N. Good question. I've been waiting all day for someone to ask me that! (explains)
Now, what about **T**?

B. It's the set containing the functions **F** and **G**.

After realizing that **S** and **T** are sets of functions, all the students, including J and B, proceeded to correctly do the remaining problems with very little help from me. Learning had occurred --- and I came to see my role as a teacher not as someone who gives forth knowledge, but as someone who is a moderator of cognitive struggling. Moreover, almost everyone in the class claimed that there were two possible answers to 1.1 (b) since sets are unordered. Their previous work with ISETL had helped.

Goal 1.2. Think of a set as an object (first pass).

Question. If students can think of one set as a legal element in another set, or even in a tuple, does this mean that they can think of a set itself as an object and not just as a collection of objects?

Problem #1.2 (done as homework, not discussed during the interviews)
Consider the following set:

$$S := \{1.2, 2+4, \{1,3..7\}, \text{"hi there"}, \text{true impl false}, [\text{"a"},\{10,9..0\},-9.8]\};$$

(a) How many items are in **S**? What are they?

(b) What might ISETL output if you type:

S;

Understanding sets: Second level

Goal 2.1. Think of forming a set as a process.

Questions. What is going on as sets are formed? Can students think about iterating over a (domain) set?
To help answer these questions, we gave them a set former and asked them for the situation, and conversely gave them a situation and asked them for the set former.

Problem #2.1. Let **I** be a set of integers.

(a) Give the verbal description of the set **T** which is defined by the following set expression:

$$T := \{x : x \text{ in } I \mid \text{not}(x < -3 \text{ and } x \bmod 2 /=0)\};$$

(b) Suppose **S** is the set of all the elements of **I** which are divisible by 2 if they are greater than 10. Write a formal set expression for **S**.

Observations. As a result of the ISETL exercises that they had done previously, almost everyone did these two problems correctly. They had no difficulties thinking about the process of iterating over a domain set to form a new set. Three students, however, had minor difficulties translating a proposition between English and ISETL. In particular, in translating the condition in part (a) from ISETL to English, they applied DeMorgan's Law and said "**x** had to be greater than –3" (instead of greater than or equal to –3), and, in translating the condition in part (b) from English to ISETL, they did not recognize that they needed to use the logical connective, implication.

Goal 2.2. Set operations.

Question. Given two set expressions as set formers can students find their union? intersection? difference? Ask them to find the symmetric difference. Do they look at what they already have? Or do they try to find the set former from scratch?

Problem #2.2. Assume in the following Venn diagram, the square represents the set **T** from part (a), the circle represents the set **S** from part (b), and the rectangle represents the given set of integers **I** in problem 2.1.

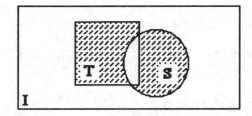

Write a formal set expression for the set expressed by the shaded portion of the diagram. Note: Do not use any set operations such as union. Instead use the conditions which are satisfied by the elements in **S** and **T**.

Observations. Four students combined the conditions for **S** and **T** using an expression equivalent to exclusive or. Interestingly, the two best students in the class arrived at a correct expression by starting from scratch and reasoning through the situation. The remaining students really struggled. Most were able to say what they wanted to do — they wanted all the items in **S** or **T** that were not in both, and consequently they wanted the conditions for **S** or **T** to hold, but not the conditions for both – but they were able to express these ideas mathematically only with a great deal of careful hand–holding. Their understanding of the logical connectives and their connections to the set operations were not strong enough to support the type of thinking that was required. Their understanding of prepositional logic needed to be strengthened.

Goal 2.3. Think of a set as an object (pass 2).

Question. Are students comfortable thinking about the fact that a function can return a set?

If so, they should be able to think of a set as an object. We decided to give them an ISETL function that returns a set and ask them what it does for various input.

Problem #2.3. What will appear on the screen after the following lines are entered in ISETL:

```
G := func (N,M);
    return {1, N..M} + {–M, –N..1};
end func;
```

```
G(2,4);         G(2,3);         G(3,2);
```

Observations. Based on class discussions and performing sample sessions such as the one given above, all but two students easily found the set values for the first two expressions. They appeared to be able to think about a set as an object returned by a function.

No one, however, found the correct answer to the third expression, **G(3,2)**. Obviously, the way that we had approached the formation of special sets of integers having the form {**p, q..r**} in class was not helping them in this situation. In class, we had noticed that **p** and **q** determine whether the set is defined by an increasing (decreasing) sequence and the size of the jump, while **r** is the upper (lower) bound. Consequently, we had derived the following algorithm: First determine whether the sequence is increasing or decreasing. What is the size of jump? What is the upper or lower bound? Now, pick the first item and test it, pick the next and test it, and so on, as long as the chosen item is less than or equal to the upper bound (greater than or equal to the lower bound). During the interview procedure we developed an alternate approach: Construct the associated sequence of integers by determining whether the set was defined by an increasing or decreasing sequence and determining the size of the jump. Then use the upper or lower bound as the cut–off point. During an early interview, one student explained it this way. I then used this explanation on subsequent students, and they understood!

Understanding sets: Third level

Goal 3.1. Doing hard (cognitively nontrivial) problems involving sets.

Questions. How do they go about doing a "hard" problem? What are they thinking about as they try to solve it?

Note: The following problems were done as homework, not during the interview.

Problem #3.1. Show that **A*B = B** iff **B** subset **A**.

(a) Do by elements.

(b) Do by manipulating formulas.

Problem #3.2. Find #(S1*S2) and #(S2–S1) in terms of #S1, #S2, #(S1+S2).

Problem #3.3. Recalling that **npow (n,S)** returns the set of all subsets of the set **S** with cardinality **n**, list the items in

npow (2, pow({"able", {1..4}})).

CONCLUSION

The results of this research process influenced the presentation of materials in the textbook *Learning Discrete Mathematics with ISETL*, by Baxter, Dubinsky and Levin (1989). Although undertaking the testing/interview process and analyzing the results is very time consuming, we have been very encouraged by the results of using materials based on this approach in the classroom (see Baxter and Dubinsky, in press). Students find that they make sense. In addition to developing instructional materials for discrete mathematics, the approach is being used to develop learner–centered activities in other areas in mathematics and computer science (see Baxter, 1991; Leron & Dubinsky, 1993; Dubinsky and Schwingendorff, 1990).

Bonsangue presents a carefully designed observation and statistical analysis of the performance of two classes in an introductory statistics course at a large public community college. One class was taught in a traditional style with lectures three times per week. The experimental class had two lectures per week, and the third session was highly interactive. Students worked cooperatively in groups of 3 or 4 on assignments directly related to the concepts in the course. The groups were initially self-selected and then maintained through the semester. Students were encouraged to use the groups in other aspects of the course such as doing homework and studying for exams.

There was no appreciable difference between the two groups in performance on homework and scores in the first exam. In the remaining three exams and in the final, however, the cooperative group definitely outperformed the traditional group and the differences grew as the semester proceeds. The differences in performance are presented, not only as straight differences, but also adjusted in terms of a predicted performance index for each student.

Every effort is made to show that, except for this difference in treatment, the two classes were identical in several respects such as previous grades in mathematics courses and in all courses. Thus we are led towards the conclusion that the element of cooperation resulted in better learning.

This paper may be considered as a data point that relates to a number of issues in research in undergraduate mathematics education. Just as workers in this area need to make choices between qualitative and quantitative paradigms in designing their investigations into how people learn, Bonsangue's paper reminds us that we must also ask the related question of how to measure and compare this learning.

The paper does not tell us about the exam questions, but we do have all the scores. On all five exams, the control group had an overall average of about 74% and the experimental groups averaged about 83%. This is a noticeable difference, and it also turns out to be statistically significant. But what does it mean in terms of learning? What can we conclude from an average improvement of 9 points? Are the cooperative students more careful and so tend to lose fewer points because of careless errors? Or is there, on each exam, a single problem which stumps one group and not the other? Or, what would be most interesting, can we show from analyzing the test questions themselves (say, in the spirit of the papers by Hart or Selden et al) that some improvement in understanding may be occurring? Hopefully, questions like these will be addressed in future studies.

Another issue arises just from looking at the exam scores. How important is it to develop teaching methods that will improve test scores from the middle seventies to the low eighties? This is certainly not, on the face of it, a situation in which students are not succeeding with traditional methods (as with some other subjects such as calculus or abstract algebra, or even more advanced statistics courses) -- and a situation in which a pedagogical innovation takes a big step towards solving this serious problem. Rather, what Bonsangue has given us is an absolutely solid piece of evidence supporting the assertion that cooperative learning is a tool that can lead to improvement and might be used to solve more serious problems.

This is particularly important in light of the information which Bonsangue gives us to the effect that some other studies of the effectiveness of the collaborative model have shown mixed results ranging from positive to negative. This is even the case with the present study. The cooperative group had an attendance record that was only slightly better and the attrition in that group was slightly worse than in the control group. In neither case were the differences statistically significant. Obviously cooperative learning, like most pedagogical techniques is not a panacea and, although it can be very effective, may not also help. An important question for research is to try to understand the conditions in which various teaching methods can make significant improvements in student learning.

Bonsangue takes a number of steps in an important direction when he goes beyond the numerical information and tries to understand the nature of students' cooperative efforts, to analyze how the groups functioned, and to speculate on what specific aspects of cooperation led to the superior performance. This is a beginning. It is to be hoped that researchers in undergraduate mathematics- education will develop an experiential base together with theoretical perspectives that can be used to make sense of the very exciting idea of cooperative learning and to point to solutions of some of the major problems of undergraduate mathematics-education.

SYMBIOTIC EFFECTS OF COLLABORATION, VERBALIZATION, AND COGNITION IN ELEMENTARY STATISTICS

Martin V. Bonsangue
Department of Mathematics
California State University
Fullerton, California 92634

ABSTRACT

The effect of a collaborative learning model on student achievement was examined for two groups of students enrolled in introductory statistics. One group used a collaborative model as part of the instruction, and the other used a traditional lecture format. External factors such as course instructor, time of class, and evaluation of student performance were controlled for both groups. Moreover, the control and treatment groups were selected at random before the course commenced. No differences were found between groups in attendance, homework completion, or course completion rates. Comparison of control and treatment groups showed no difference in achievement on the first exam; however, significant differences favoring the treatment group were observed at all measurement points thereafter. Time series residual analysis was performed to illustrate the widening gap of achievement effects throughout the semester once preintervention effects were removed. Students' understanding of statistical concepts and applications are discussed, together with an example of a student conversation. Implications for alternative classroom structures in college mathematics courses are considered, together with the need for replication and further study.

INTRODUCTION

The use of collaborative learning in higher education has attracted the attention of researchers and educators in a variety of disciplines (Bruffee, 1987; Mangan, 1987; Wiener, 1986). Academic networks such as the Learning Assessment Retention Consortium (LARC), in which college instructors employ a collaborative model as a primary structure of the course, have generated interest in the efficacy of collaborative learning across the curriculum (Lohnas, 1990). Although some academic disciplines may inherently be more interactive than others, intentionally structured groups are typically employed in this model as an alternative to the traditional lecture/listen format as the sole medium of teaching. The use of the collaborative model in mathematics is particularly interesting in light of recent trends in mathematics-education to enhance student interest and learning with current technological tools and pedagogical approaches (Steen, 1987; Treisman, 1985).

Studies of the effectiveness of the collaborative model upon mathematics achievement at the primary and secondary levels have shown mixed results. Some researchers have reported positive achievement effects (Hamblin, Hathaway, and Wodarski, 1971; Slavin and Karweit, 1982), while others found no differences (Peterson and Janicki, 1979; Slavin and Karweit, 1981) or even negative effects (Johnson, Johnson, and Scott, 1978). Nonetheless, instructors of mathematics in colleges and universities have of late shown increased interest in the use of interactive student groups, particularly in the teaching of calculus (Barrett and Browder, 1989; Bonsangue, 1992; Steen, 1987; Treisman, 1985). However, the literature has not included efficacy studies of interactive learning for courses in elementary statistics. Statistics is a prerequisite for most business and math–based science majors, and remains for some students a stumbling block to their academic or programmatic advancement. The present study examines the effect of collaborative learning upon student achievement in an introductory statistics course. Careful analysis of the effectiveness of the collaborative model upon student achievement and learning in this study may help illuminate the feasibility of adapting a collaborative structure in other post–secondary mathematics courses.

The present study examines the interdependence of cognition with verbalization through interaction with others. This interdependence, or symbiosis, provides the structure upon which individuals create language and metaphor, permitting them to understand a new and complex idea (Pimm, 1987). Although difficult to define and measure directly, there is evidence in the literature on college and university mathematics that student interaction in academic settings produces measurable effects upon learning and achievement (e.g., Treisman, 1985). This study reports the grade performance for statistics students in a collaborative setting, together with observations of the learning process in group settings.

METHOD

SUBJECTS

Subjects were students enrolled at a large public community college in southern California. Two sections of introductory statistics classes taught in the mathematics department for the 17–week fall semester, 1990, were targeted as the sample groups. Both were morning sessions (50 minutes MWF) taught by a full–time member of the mathematics department involved in the LARC project. Prior to student registration, one section was targeted as the control (traditional lecture) group, and the other as the treatment (cooperative learning) group. On the first day of class, students in each group were given a written syllabus describing grading policies and related course information. Both outlines were identical except that the cooperative group syllabus included a description of how interactive learning would be an integral structure of the class sessions throughout the semester. Although the cooperative group students were given the opportunity to transfer to a traditional lecture course, none chose to do so.

The content of the course centered on topics typically listed in course descriptions for elementary statistics, including probability and probability distributions, sampling and research design, descriptive and inferential statistics, correlation and regression, and nonparametric statistics. The text used was **Elementary Statistics,** Fourth Edition, by Mario Triola (1989). Students used calculators throughout the course; there was no computer component used in the course.

COURSE COMPLETION

The cooperative group began the semester with 38 students (17 male, 21 female) while the control group initially had 30 (9 male, 21 female). Seventy–six percent (13/17) of the men and 67% of the women (14/21) completed the course in the cooperative group, while 89% (8/9) of the men and 71% (15/21) of the women completed the course in the control group.

By ethnicity, 75% (12/16) of the minority students and 68% (15/22) of the non–minority students in the cooperative group completed the course. Completion rates for the control group were 78% (7/9) for minority students and 76% (16/21) for non–minority students. Table 1 shows course completion breakdowns by gender and by specific ethnic groups.

Comparison of completion rates for various categories were tested for statistical significance. Chi-square analyses revealed no differences between men and women within the control group ($X^2 = 1.55$, $p > .15$, $df = 1$) or within the cooperative group ($X^2 = 1.16$, $p > .25$, $df = 1$). Likewise, comparison of completion rates for both men and women between groups showed no difference ($X^2 = 1.17$, $p > .25$, $df = 1$). Collapsing ethnic groups into minority (non–white)/non–minority (white) status likewise yielded no differences within the control group ($X^2 = .012$, $p > .30$, $df = 1$) or the cooperative group ($X^2 = 0.54$, $p > .30$, $df = 1$). Since data was collected only on those students who completed the course, the present study reports results based on those students. Possible effects of the cooperative model upon persistence are considered in the discussion section.

INSTRUMENTS

Grades in each group were based on four 50-minute teacher–constructed written examinations (in weeks 3, 7, 11, and 15), a 2.5-hour final, and homework assignments. Exams and the final accounted for 85% of the grade, with the other 15% based on homework completion. Both groups were given the same examinations, and standards of grading were identical for both groups. Although class attendance did not directly affect the students' grades for either group, attendance records were kept for each student throughout the semester.

Table 1
Course completion for cooperative and control groups by gender and ethnicity

	Cooperative Group			Control Group	
	Week 1	Week 18		Week 1	Week 18
By gender					
Male	17 (45%)	13 (48%)		9 (30%)	8 (35%)
Female	21 (55%)	14 (52%)		21 (70%)	15 (65%)
By ethnic					
Asian	2 (5%)	1 (4%)		2 (6%)	2 (4%)
Black	5 (13%)	3 (11%)		2 (6%)	1 (4%)
Hispanic	9 (24%)	8 (30%)		5 (17%)	4 (17%)
White	22 (58%)	15 (55%)		21 (70%)	16 (70%)
Total	38	27		30	23

CLASSROOM STRUCTURE

The cooperative class and the control class were taught at different times in the same classroom. Lectures by the instructor were given three times per week to the control group and twice a week to the cooperative group. The third session was highly interactive. Students worked together in self–selected groups of three or four on an assignment directly related to the concepts presented that week in the lectures and homework. Cooperative lessons were highly structured by the instructor with clear written directions and goals. Students were free, however, to work together with other groups on the lesson. The in–class assignment was handed in by each group, and was included as part of the students' homework assignment.

COGNITIVE CONTROLS

Since the control and treatment groups were selected randomly before registration began, it was reasonable to expect that there would be no difference in mathematics ability or previous success between groups. In order to test this hypothesis, the groups were compared on two variables that have been shown to be strong predictors of mathematics success: previous mathematics grade and overall college grade-point-average (Dew, Galassi, and Galassi, 1984; Llabre and Suarez, 1985; Siegel, Galassi, and Ware, 1985). The students' grades in their previous college mathematics courses together with all–college GPAs were retrieved from institutional records on those students for whom such data was available. Because of the wide variety of students at a community college, no standardized measures such as the SAT or ACT existed for more than a few students. Moreover, there was no placement exam required for admittance into the statistics course; students needed to have passed a course in intermediate algebra with a grade of C– or better.

RESULTS

ANALYSES OF CONTROL VARIABLES

T–test analyses for independent samples showed no differences between the cooperative and control groups on previous mathematics grade ($t = 1.39$, $p > .15$, $df = 34$) or GPA ($t = -.97$, $p > .30$, $df = 48$). Thus neither the cooperative nor the control group of students seemed to have an inherent mathematical or scholastic advantage. Furthermore, there was no significant difference in class attendance throughout the semester between the groups. The cooperative group averaged 3.4 absences out of the 51 meetings, while the control had 3.6 absences group ($t = .76$, $p > .20$, $df = 48$), suggesting that neither group received more direct instruction.

EXAM SCORES AND COURSE GRADES

Exam scores, reported as percents, were tracked throughout the semester to determine when and whether group differences between the cooperative and control classes might appear. On the first exam there were no differences in achievement, with the cooperative class receiving a mean score of 79.3, and the control group a score of 80.0. Differences in homework scores for the semester were also nonsignificant, with students who stayed in the course successfully completing the homework assignments. However, significant differences between groups on exam scores were observed for exam 2, exam 3, exam 4, and the final exam. The differences increased during this time, with the cooperative group outscoring the control group by about 10 percentage points on exam 2 and exam 3, 12 points on exam 4, and 16 points on the final. Table 2 reports the mean homework and exam scores and standard deviations for each group throughout the semester. It is important to note that these exam score averages include only those students completing the course, not for all students who took the exam.

Course grade comparisons between the cooperative and control classes showed similar results since course grade was largely based on the exams and final. Grade distributions for each group were as follows: the cooperative group received 9 A's (33%), 17 B's (63%), and 1 D (4%), while the control group received 5 A's (22%), 9 B's (39%), 6 C's (26%), 2 D's (9%), and 1 F (4%). The difference in mean grade for the control (2.65) and the cooperative (3.26) groups was significant at the .05 level (Table 2).

Table 2
Mean homework, exam, final, and course grades for cooperative and control groups

	Home–work	Exam 1	Exam 2	Exam 3	Exam 4	Final Exam	Course grade
Cooper.	100.0	79.3	82.7	82.0	89.3	84.0	3.26
$n = 27$	(0.0)	(11.3)	(13.7)	(13.3)	(13.0)	(9.8)	(0.66)
Control	98.6	80.0	72.7	72.7	77.3	68.0	2.65
$n = 23$	(6.9)	(13.0)	(18,7)	(21.7)	(20.2)	(20.9)	(1.07)
t–value	1.00	–.07	2.16*	1.82*	2.45*	3.36#	2.37*
($df = 28$)							

Note. Exam scores are reported as percentages; standard deviations are given in parentheses.

* $p < .05$ # $p < .01$

It is difficult to ascertain whether achievement effects were felt more strongly by any gender or ethnic subgroup. As previously indicated, no differences in course persistence were observed between specific groups. Likewise, comparison between minority versus non–minority status for combined classes showed no differences for course grade ($t = 0.13$, $p > .9$, $df = 48$). Controlling for minority status between groups showed mixed results: minority students showed no differences ($t = -.51$, $p > .6$, $df = 17$) while non–minority students showed extreme differences ($t = 3.78$, $p < .001$, $df = 29$). It is possible, however, that this effect is explained by gender more than by minority status. Controlling for gender between groups showed that, although there were marginal group differences among the men ($t = 1.65$, $p < .15$, $df = 19$), there were more pronounced differences among the women ($t = 2.43$, $p < .05$, $df = 27$). As previously mentioned, control and treatment groups were randomly determined before the class sessions began. The imbalance between the numbers of men and women (8 and 15, respectively) in the control group, coupled with the fact that 12 of the 15 women in the control group were white, indicates that caution must be exercised in interpreting inferences about specific ethnic or gender groups.

TIMES SERIES RESIDUAL ANALYSIS

The results presented above are based on data gathered on both treatment and control groups over successive time intervals. Drew (1983) has described a technique of graphically presenting time series data for which "preintervention data on important predictors of the criterion" are available (p. 97). In this technique, variables which are powerful predictors of the outcome measure are used to generate a separate regression model for each time-point. A residual is then obtained for each student for each time-point. This yields for each individual "an outcome measure … from which the effects of known preintervention covariates have been removed" (Drew, 1983, p. 97). The mean residuals for both the treatment and control groups are then plotted, providing a graphical representation of differences between the two groups on the outcome variable over time. These differences can then be tested for significance using standard t or F tests (Drew, 1983). This technique is helpful for the present analysis since it provides pictorial representation of the direction of the changes of residual exam scores between the cooperative and control groups at points in time throughout the semester.

As discussed earlier, previous mathematics achievement and college GPA have been shown to be powerful predictors of mathematics success. These two variables were used as the preintervention measures to predict scores on each exam and the final. Separate regressions based upon the entire sample for whom preintervention data were available were run for each exam, yielding a multiple correlation coefficient at each time point. The standardized regression coefficients (beta weights) were also computed. Using these beta weights as the coefficients, a predicted exam score was calculated for each student in the control and cooperative group. The actual exam score was subtracted from the predicted score, yielding a residual score for each student at each time point.

These residuals were then averaged for the control group and cooperative group separately, giving a mean residual of predicted versus actual achievement in which preintervention achievement factors have been taken out. For example, on exam 2, previous math achievement and college GPA produced a multiple R of .64; thus these two variables explained about 40% of the variance of the scores on that exam. The beta weights of .37 and .34 were used to calculate a predicted exam score for each student. Students in the cooperative group averaged 3.9 percentage points above their expected score, while those in the control group averaged 3.5 points below. Table 3 reports the multiple correlation coefficients, regression coefficients, and mean residuals for each group.

Table 3
Residual analysis of mean exam and final scores

	Exam 1	Exam 2	Exam 3	Exam 4	Final
Multiple correlation coeffs.					
(R)	.36	.64	.49	.41	.41
F	2.45*	11.12#	5.29*	3.32*	3.24*
Standardized regress. coeffs.					
GPA	.28	.34	.28	–.11	.05
Previous grade	.11	.37	.28	.46	.37
Mean residuals (in %)					
Cooperative group (n=17)	–0.9	3.9	5.1	5.5	7.1
Control group (n=19)	0.9	–3.5	–4.5	–4.9	–6.4
Residual t–test	–0.52	1.76	1.87*	1.96*	2.58*

$* p < .05$ $\# p < .01$

Figure 1 shows the graph of the mean exam scores for the cooperative and control classes at each time-point. The scores of the two groups are roughly parallel to each other after exam 2. However, after removing the preintervention effects, we get a clearer sense of the trends in achievement throughout the course of each group. The graph of the mean residuals (Figure 2) suggests that the cooperative group achieved higher than preintervention indicators would predict, and that the control group achieved lower than would be predicted. Moreover, this gap widened throughout the semester, culminating in a significant residual difference between the two groups on the final exam ($t = 2.58$, $p < .05$, $df = 34$).

Figure 1
Exam and final scores for cooperative and control groups.

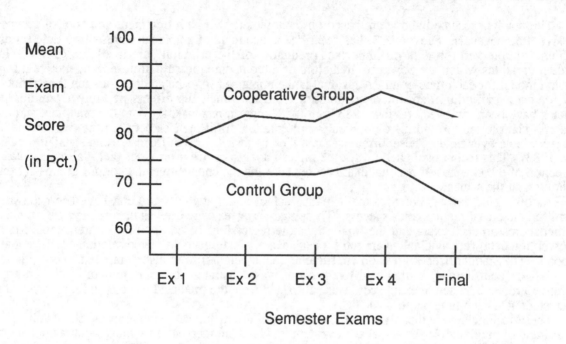

Figure 2
Residual exam and final scores for cooperative and control groups.

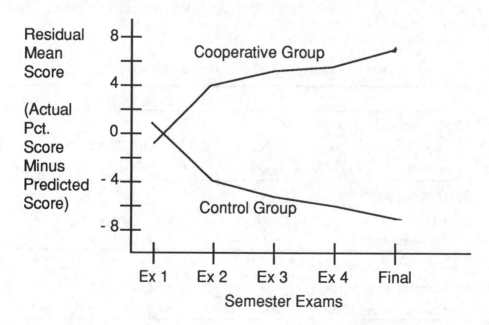

DISCUSSION

The time series residual analysis suggests that, after preintervention factors were removed, the collaborative model seemed to increasingly affect the achievement of students in the cooperative class throughout the semester. Those students in each group who completed the course scored nearly identically on the first exam, but produced a gap in achievement of more than 3 standard deviations on the final. Stated differently, the use of the cooperative structure led to better than acceptable or predicted results, while its non–use led to less than acceptable results.

The connection that students in the cooperative group made among themselves and their colleagues and the instructor may have increased their commitment to the course and their mathematical self–confidence. The course instructor observed that after the first exam — which yielded the lowest mean score among the four exams for the cooperative group — students in the cooperative class began working together outside of class as well as during class. Study groups including the majority of the cooperative students met together in the math lab at least twice a week to discuss and solve homework problems. Many students from the cooperative class also came to the instructor's office hours regularly as a group, whereas students from the control class came individually if at all. The instructor observed students in the cooperative class to be generally better prepared in class, more attentive during lectures, and more interested in the subject.

A STATISTICAL EXAMPLE

The difficulties experienced by many students in an introductory statistics course typically lie in confusion about the deeper concepts underlying the mechanical processes. There is evidence to suggest that verbal interaction among students in mathematics courses helps to clarify their understanding of the mathematical principles involved as they construct metaphors that are meaningful for them (Pimm, 1987; Treisman, 1985). As mentioned previously, the instructor overheard many conversations about statistics during the interactive work sessions. Two common questions seemed to emerge rather frequently for most of the students having trouble:

1. How do you know when to use which formula?

2. Why is the formula built as it is, and what does the numerical result tell us??

The following account reconstructs a fairly typical conversation overheard during the collaborative sessions, here among three students working problem 6–79, page 370, in *Triola*. The instructor had not yet lectured on this section.

> In a study of the long–term effects of promoting or holding back elementary school students, the following reading test results were obtained for a sample of 15 third-grade students: $m = 31.0$, $s = 10.5$. (The data are based on "A Longitudinal Study of the Effects of Retention/Promotion on Academic Achievement," by Peterson and others, *American Educational Research Journal*, Vol. 24, No. 1). Does this third-grade sample mean differ significantly from a first grade population mean of 41.9? Assume a 0.01 level of significance.

Student A: So we're comparing the sample mean of 31 against the population mean of 41.9. Which formula do we use?

Student B: Probably the one in this section. What else would you use?

Student A: OK, so we have to set up a null hypothesis. Would it be a one–tail or two–tail test?

Student C: Two tail, because it didn't say it had to be less.

A: But it *is* less!

C: But that could have happened due to chance in the sampling. Better use a two–tail.

A: (After computing statistic) I think we should fail to reject the null hypothesis, because it's not out far enough.

C: What'd you get?

A: Negative 1.04.

C: Did you divide by $\sqrt{n}$?

A: No. We didn't before.

C: But we have to now.

A: Why?

B: 'Cause it's in the section. I got negative 4.02.

C: Me, too, and it's in the back of the book. So reject the null hypothesis, since it's past the mark.

A: What would we use for a significance level if they didn't tell us?

C: I dunno. Point oh five, I guess.

A: But that means it would be easier to be significant, then, right?

B: I guess so, but we got the right answer. We're done.

     This dialogue shows examples of both correct and incorrect thinking and statistical usage. Student A gives a correct interpretation to her result, even though it came from the wrong formula. Although the other two students got the correct numerical answer, none of the three really understand why $\sqrt{n}$ needs to be in the formula, other than its prominent placement in the section. Although the language is not quite right, ("reject the null hypothesis, since it's past the mark"), the concept of inference based on probabilistic occurrences is, at least in part, clear.

     The instructor was able to talk about the problem with the group, and help clarify why the $\sqrt{n}$ is used when testing a sample mean. In fact, it was only after wrestling with these problems in Chapter 6 that the concept of standard error of the mean from the previous chapter had any relevance or importance. Indeed, this concept was discussed with the class in lecture during the following session. Students in the cooperative class were much more successful than students in the control on the unit exam and final exam in which several different types of inference testing problems were asked. Having no "section" clues, students had to understand why one formula would be used over another in a given situation. If the instructor had not had an opportunity to hear this dialogue, the specific questions and ideas that it generated would probably have never surfaced.

GENERALIZATIONS AND LIMITATIONS

     There are limitations to the generalizability of this type of research. The sample size is limited initially to the number of students enrolled in two sections of the course taught by the same instructor, and decreases by the end of the semester. During this time there is self–selection in the attrition that occurs, which is largely outside of the control of the instructor. It may be that students who do not like the cooperative structure silently slipped away, leaving a somewhat skewed sample of individuals more comfortable in interactive settings. Also, the heterogeneity of student experience and attitudes not explained by previous academic or mathematics achievement typical at a community college makes it nearly impossible to identify or control for differences in affective factors. The lack of use of technology may be important as well, since students working at computers, usually in teams of two, will interact naturally and often be more motivated in their learning. Nonetheless, this study found evidence that interactive learning is not only correlated with, but can effect, positive changes in mathematics achievement.

     It is interesting that the differences in group achievement occurred fairly consistently over time after starting the course "even." It may be a question of semantics whether or not the control group fell behind or the cooperative group got ahead. Although the grade distribution in the cooperative class seemed to be higher than that for statistics courses taught by the same instructor in previous semesters, this was not analyzed statistically. Indeed, even with significant differences it would be difficult to make general inferences, due to the small sample size and lack of controls for the previous groups. The key factor seems to be that students in the cooperative group spent more "time on task" in class, at school, and at home, talking about and doing statistics, than did students in the non–interactive structure. It is tricky to gather hard data for constructs such as "time on task," particularly in college courses. However, anecdotal data based on student observation supports the somewhat obvious contention that those who spend more time at a task generally do it better than those who spend less time.

     The observations described above are consistent with the literature on the powerful interplay between socio–academic factors and cognitive development. Pascarella (1985), in a thorough review of the literature, has posited a causal model "for assessing the effects of differential environments on student learning and cognitive development" (p. 50). In the model, interactions with faculty and peers appear as an intervening wave of variables directly affecting cognitive development. Moreover, these effects may be very strong for students attending commuter institutions (Astin, 1968; Tinto, 1987).

CONCLUSION

The present study found positive effects in course achievement in statistics which may be connected to the interactive nature of the course. The weekly cooperative learning sessions seem to have served as a catalyst by which students came to interact together on the assignments outside of class. Indeed, the relative success of the cooperative group may be in part explained by the effectiveness of the instructor in a cooperative structure as well as by the individual or collective efforts of the students. Using an interactive structure during class-time removed some of the control of the pace of the course from the instructor; consequently, time-management in selecting lecture topics and generating meaningful group sessions with specific problems was crucial. However, during these sessions, the instructor had the opportunity to listen to students' correct or incorrect thinking, and to address these problems individually or collectively.

The effect of the collaborative model upon the students not completing the course is difficult to ascertain without gathering follow–up data from those individuals. Although completion rates were not significantly different for the cooperative and control groups, there is evidence to suggest that a disparity between a student's college experience and that student's expectation may impact the individual's persistence in that situation (Tinto, 1987). Clearly, replication is needed for further corroboration of the results presented in statistics as well as for other pre–baccalaureate level mathematics courses, although the necessity for strict comparison controls makes larger samples of specific subgroups difficult to obtain. In summary, this study found evidence to suggest that use of a cooperative learning structure may be an effective mode of teaching and learning mathematics in higher education.

The research reported in this paper is according to a paradigm completely different from the controlled experiment and statistical analysis that we saw in the paper of Bonsangue. Cuoco begins with a description of the theoretical perspective which frames his research. Then he gives some detailed examples of student work. This is followed by assertions (which he calls conclusions) that relate his theoretical analysis to his observations of students and provide pointers to pedagogical practice. In these two papers we have (very roughly) contrasting examples of quantitative and qualitative research.

Cuoco is concerned, as are many researchers, with process and object conceptions of functions. He argues that both conceptions are necessary in the school curriculum that leads up to Calculus and considers the extent to which students (and their teachers) at this level have such conceptions. He concludes that many have neither, and discusses pedagogical approaches, using computers, that could help develop both process and object conceptions.

A major concern of Cuoco is what he calls the "$\mathbf{R}$ to $\mathbf{R}$ curriculum". This refers to his characterization of the standard precalculus curriculum as a study of the properties of what he calls a *higher-order function* $G{:}\mathbf{R}^{\mathbf{R}} \to P(\mathbf{R}^2)$. That is, G acts on a function f from $\mathbf{R}$ to $\mathbf{R}$ and produces a subset of $\mathbf{R}^2$ — the graph of f. He shows, by giving several examples, how a large part of the curriculum involves a study of the properties of G in the sense of describing the effect of changes in f on G. It is interesting to see how these changes can be specified formally in terms of G.

Cuoco argues that understanding this curriculum requires students to understand functions as objects. This is possible, he asserts, if students have a conception of a function as a process. In that case, the need to understand G can lead to their encapsulating these processes into objects, and the curriculum will make sense. Unfortunately, he points out, all too few students come to precalculus with an understanding of functions as processes. As a result, they often change the activities and are reduced to learning about the connection between equations and graphs, which is not enough.

It is an intriguing analysis, and it leads Cuoco to suggest a very different curriculum which has as its main goal that students first construct a conception of a function as a process and then learn to encapsulate these processes so that functions may be seen as objects which can themselves be transformed. He argues that this can be done effectively by having students make constructions on computers using such software as Function "Machines", Logo, and ISETL.

The paper provides a number of examples of how Cuoco has used the programming languages Logo and ISETL to help both students and teachers enhance their understanding of functions. His use of ISETL may be compared with Baxter's paper on the development of the set concept. With Cuoco's work, as with that of most other researchers, the object conception is seen to be much more difficult for students than is the process conception.

Cuoco's paper raises two kinds of curricular questions. First, given the present K–12 curriculum, what sort of conceptions of function are really necessary for students? And second, are there alternative curricular approaches that might change the answer?

It is very encouraging that, in suggesting some future directions that work like this might take, Cuoco is optimistic about how far students can go in developing sophisticated understandings of abstract mathematical concepts such as functions. Research in Mathematics Education began, of necessity, with discovering, pointing out, and analyzing the profound difficulties that students have with understanding mathematics. Eventually, our research should reach a point at which we can begin to propose, implement, and report on remedies. Cuoco's paper suggests we may be approaching that point, at least in some areas.

MULTIPLE REPRESENTATIONS FOR FUNCTIONS

Albert A. Cuoco[5]
Woburn High School and Education Development Center

ABSTRACT.

What mental models for the function concept are required to deal with the current precalculus and calculus curriculum? In this paper I provide examples that show that the curriculum requires both a process and an object concept for functions (as defined in Breidenbach et al, 1992) but that it doesn't provide any activities that encourage students to develop these concepts.

Building alternate procedural descriptions on a computer for a given function can help high-school students develop a process concept for functions. Constructing arguments that show that two processes describe the same function helps students view functions as objects. Building models in a computer-language that allows for functions as first-class objects can help some students develop a general object concept of functions.

These conclusions are based on work with four groups of students: three classes of high-school students at Woburn High School in Woburn, Massachusetts; and a class of thirty high school teachers who spent two summers at the Institute for Secondary Mathematics and Computer Science Education (IFSMACSE) at Kent State University in a course that centers around the construction and application of computer models for functions.

INTRODUCTION

The use of complex objects as single entities while preserving their non–atomic nature is a distinguishing feature of modern mathematics.

In elementary mathematics, the whole process of comparing one integer a to another integer b by taking their quotient is expressed as a single *number*, $\frac{a}{b}$, and students are expected in the curriculum to move between treating $\frac{a}{b}$ as an object (when they manipulate fractions, for example) and as a comparison between integers.

In advanced mathematics, consider measure theory. An important idea here is that of a *set algebra*: Starting with a set S, certain subsets are gathered into a set and are manipulated as objects. The fact that these objects are themselves sets (subsets of S) is never suppressed. The definition and properties of a set algebra depend directly on the nature of its elements, so that its elements are viewed at once as sets and as elements of other sets.

As another example (closer to the topic of this paper), consider the case of Galois theory: Certain kinds of mappings on a field are collected up into a set; this set is given the structure of a group; and then transformations of this group (*another* collection of mappings) are studied in order to yield information about the structure of the underlying field. The fact that the elements of the group are mappings is never suppressed (many of the structural properties of the group are derived by making explicit reference to the nature of its elements), so that the mappings on the underlying field are viewed at once as *processes* (functions that act on the elements of the field) and as *objects* (elements of a group that can be transformed and combined).

The words "process" and "object" in the previous paragraph are part of the vocabulary used in the extensive body of research on how people understand functions. Dubinsky and Harel (1992) use them to describe the top end of a hierarchy of levels of understanding that has been observed in many students. Anna Sfard (1992) uses the adjectives *operational* and *structural* to describe similar levels. Other authors use words like "dynamic" and "static" in the same spirit.

---

[5]Work supported by NSF grant MDR-8954647. Many of the ideas in this paper arose from conversations with Jim Brennan, Paul Goldenberg, Phil Lewis, and Jim O'Keefe. Special thanks to Don Muench, who presented a draft of this paper at the special session for undergraduate research at the San Francisco joint AMS-MAA meeting in January, 1991.

This technique of using compound objects as simple data has become widespread in mathematics because it has proven to be enormously powerful. One of the reasons it interests educational researchers is that, in many ways, it provides a concrete model for a theory of epistemology (due to Piaget) based on the notion that new knowledge is constructed via a recursive process of *interiorization*, where a set of actions on objects is recognized as a repeatable process, followed by *encapsulation*, where that process is transformed into a cognitive object that can be acted upon by higher–order manipulations. When the mathematics involves using functions as both processes and objects, the correspondence between the mathematical activity and the epistemological perspective is almost perfect; it is little wonder that the different levels of understanding that students develop about functions hold such interest for educational research.

Why is it that so many mathematics students are unable to adopt this mathematical point of view that requires phenomena to be used on more than one level? In this paper I take a look at the well studied case of functions. First, I'll summarize an epistemological perspective that describes the development of the function concept in people who exhibit a flexible and general notion of function in their mathematical work. Next, I'll compare this perspective with the standard mathematics curriculum leading up to calculus (the "**R** to **R** curriculum"), showing how, in its myopic attempt to exploit the connection between the analytic behavior of a function from **R** to **R** and the geometric attributes of its Cartesian graph, this curriculum ignores large pieces of the developmental process necessary for students to reason about functions in general (and *even* about the very **R** to **R** functions that are the focus of the curriculum). Finally, I'll give several anecdotes that outline some alternatives to the **R** to **R** curriculum. These alternatives are based on developing flexible and usable concepts of function as *processor,* using computer models and visualizations other than Cartesian graphs. Evidence will be provided that activities of this sort may even be useful in helping people move to the stage where they can manipulate functions as objects.

THE THEORETICAL FRAMEWORK

Several authors (Dubinsky and Sfard, for example) have applied the constructivist perspective to the acquisition of the function concept. These researchers have identified several ways that students look at functions, and they have observed that each of these approaches to the notion of function can exist at several levels of abstraction. Using the language of (Breidenbach, et al, the three mental constructs of interest here are the *action*, *process*, and *object* concepts of function.

Students who view functions as *actions* think of a function as a sequence of isolated calculations or manipulations. Asked to describe a specific function ($x \mapsto x^2 + 1$, complex conjugation, or the derivative, for example), they will often reply with the last instance of the function's *application* ("$3^2 + 1$" or "the derivative of sin is cos," for example). Students holding the action concept of function think of, say, the function $x \mapsto 3x + 2$ as a multiplication by 3 and an addition of 2. The image of 4 under this function is often calculated like this:

$$4 \times 3 = 12$$

$$12 + 2 = 14$$

The first frame of Figure 1 gives a mechanical picture of the action concept for $x \mapsto 3x + 2$. Because the action concept of function is so tied up with explicit calculations, each output of a particular function is the result of an empirical proposition. This point of view makes it hard for students to deal with functions whose outputs are difficult to calculate (the sine function, for example).

Students who view functions as *processes* think of functions as dynamic (single–valued) transformations that can be composed with other transformations. They describe $x \mapsto 3x + 2$ by saying something like "it triples and adds 2," and the derivative is described as "the limit as Δx approaches zero of ..." Rather than viewing a function as a sequence of isolated calculations, students who think of functions as processes think of a repeatable *network* of calculations that can be connected to or embedded in another network, that can be reversed, or that can be iterated. Calculation of the image of 4 under x:$\mapsto 3x+2$ often looks like this:

$$4 \times 3 + 2 = 14$$

The second frame of Figure 1 shows a network image for the function $x \mapsto 3x + 2$. Familiarity with a specific function (through its repeated use, for example) causes students to *interiorize* the network of calculations, as in the third frame of Figure 1, so that each piece of the network is less important than the entire process. Such students see a similarity between $x \mapsto 3x + 2$ and $x \mapsto 5x + 1$. They are quite secure in believing that the network of delicate measurements needed to calculate $\sin 23.5°$ has been faithfully incorporated into their calculator, and, rather than agreeing with the results of an empirical experiment, the standard for correctness is taken as agreement with a rule (say, pressing the sin key). Questions about the *procedural* nature of functions ("Is it one–to–one?" "Is it an involution?") make sense for process-oriented students.

When students are able to test two functions for equality, when they can manipulate functions, when they speak of *a* conjugation or *an* absolute value (abstracting off the essential features of complex conjugation on **C** or of the usual archimedean absolute value on **R**), they are viewing functions as *objects*. For such students, functions are first–class data — atomic structures that can be inputs and outputs for *higher–order* processes. At this advanced level of mathematical thinking, students view the derivative as a function on *functions*, and, rather than saying "the derivative of x^3 is $3x^2$," they use language like "the derivative of $x \mapsto x^3$ is $x \mapsto 3x^2$." The last frame of Figure 1 is an image of an encapsulated version of $x \mapsto 3x + 2$, where the details of the process are no longer discernible, and the function is simply a black–box piece of data whose inner workings can be examined only by opening up (or *de–encapsulating*) the outer shell.

Figure 1

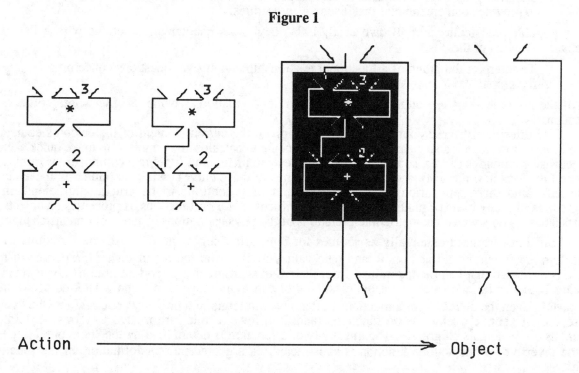

Action ————————————————————→ Object

These three ways of thinking about functions suggest a continuum of cognitive development, but, of course, things are much more complicated for students learning mathematics. How a person views a particular function depends on the particular function under consideration, how familiar it is to the person, and *how it is being used*. As in the case of Galois theory mentioned in the introduction, the most flexible use of functions comes when students view them in several ways at once. So, when studying transformations of the plane, for example, students with mathematical power see a linear mapping as a process that moves points in a particular way *and* as a matrix that can be manipulated via algebraic calculations, and the real power and beauty of the mathematics (say, the Hamilton–Cayley theorem) come from combining both points of view. In learning about group theory, many people believe that students should first study permutation groups. In such treatments, students move between viewing the permutations as objects (when they worry about commutativity, the existence of subgroups, and the like) and as processes (when they look at the actual structure of the permutations, considering parity and cycle decomposition, for example), and the two points of view are developed in tandem. One hypothesis of this theory of epistemology is that processes must act on and produce mental or physical *objects*. In order for students to understand functions defined on numbers, they must have encapsulated the notion of number; functions that produce sets are useful and meaningful only to people who can view sets as *things*. Another hypothesis is that the only way a phenomenon *becomes* an object for a student is through the encapsulation of a process. It follows that, if students are to use functions as objects (a cornerstone, as I argue below, of the **R** to **R** curriculum), they must have first dealt with these functions as processes. And, if students are to use functions as *both* processes and objects (again, an important skill in the **R** to **R** curriculum), they must be able to *de–encapsulate* functions–as–objects into the underlying processes.

Arguments are presented below to show that students come to secondary school quite capable of viewing functions as processes (see also Cuoco, 1992), and they often do so in situations where analytic properties of real valued functions of a real variable are not the objects of study. But the treatment of functions that they experience in the **R** to **R** curriculum immediately preceding and during calculus requires students to manipulate functions as objects while providing virtually no activities by which they can treat the functions as processes. Let's briefly take a look at some details of this curriculum. I'll then provide some evidence for these claims.

THE R TO R CURRICULUM

The introduction to the chapter on functions in a version of Thomas' *Calculus* (1976) states:

> Calculus is used to study the behavior of functions. Conversely, specific functions
> and their graphs are often used to illustrate facts in calculus, to suggest theorems, or
> to provide counterexamples that disprove conjectures.

A very popular precalculus text (Brown et al, 1984) begins its chapter on functions with a list of six objectives. The first of these is:

> To interpret the graph of a functional relationship and answer questions concerning
> the graph.

Five of the six objectives use either the word "graph" or "equation." None use the words "process" or "procedure."

The priorities of these courses are clear, and, given the current content of the calculus course, the source of these priorities is also clear. Students should come to calculus knowing how to make quick sketches of piecewise continuous curves, so that, in calculus, they can manipulate algebraic combinations of classical analytic functions in order to find slopes of tangent lines to these curves and areas bound by these curves. These tasks and their applications *are* important, and they continue to be crucial ingredients in the backgrounds of many users of mathematics. But their focus is *not* on functions as processes or objects; the emphasis here is on stylized geometric interpretations of the physical notions of rate, continuous change, and limit. Functions are used essentially as sources for interesting curves in $\mathbf{R}^2$. All the functions in this curriculum are functions defined on $\mathbf{R}$ and take values in $\mathbf{R}$. And the distinguished properties of these functions are those that can be determined from their Cartesian graphs; important notions about functions include whether or not they are increasing, what kind of concavity they exhibit, and what kind of symmetry they have.[6] Even the *definition* of a function is often given in terms of a property about graphs (the "vertical line test"). Algebraic operations on functions (addition, for example) are treated in terms of pointwise operations on y–heights on graphs. The injectivity of a function is determined by the "horizontal line test," and the inverse of a one–to–one function is constructed via a geometric transformation on the function's graph.

The most important function in the $\mathbf{R}$ to $\mathbf{R}$ curriculum is the one that associates a function (from $\mathbf{R}$ to $\mathbf{R}$) with its Cartesian graph, and a great deal of the curriculum is spent determining the functorial properties of this correspondence. More precisely, if f is an $\mathbf{R}$ to $\mathbf{R}$ function, let $G(f)$ denote its graph, a subset of $\mathbf{R}^2$ (identified with the Cartesian plane). So, if f is a polynomial function, $G(f)$ is a "polynomial curve" in the plane. If f is the sine function, $G(f)$ is the periodic sine wave. In general, $G(f)$ is the set of points with coordinates $(a, f(a))$, where a ranges over the domain of f. Now, the set of all $\mathbf{R}$ to $\mathbf{R}$ functions can be given many rich structural properties. It inherits algebraic structure from the algebraic structure of $\mathbf{R}$ (these functions can be added and multiplied, for example). There are also important analytic properties that such functions may or may not satisfy (continuity or differentiability, for example) that depend on the topological structure of $\mathbf{R}$. And there are function–algebraic operations (composition or inversion, for example) that make sense for functions defined on any set. A major goal of the $\mathbf{R}$ to $\mathbf{R}$ curriculum is to give geometric interpretations of these structures by seeing how they behave under G.

Many of the activities in the curriculum ask students to describe the change in a function's graph produced by a change in the function or to describe the change in a function produced by a change in its graph. For example:

---

[6] These properties all depend on delicate topological properties of $\mathbf{R}$ that are seldom motivated (or even mentioned) in introductory courses.

- Activities accompanying graphing software often ask students to investigate the change in a linear or quadratic function's graph brought about by changes in the coefficients that define the function.

- The following problem is taken from (Brown , 1984, p 277):

> In exercises 5 and 6, you are given the graph of $y = f(x)$. Sketch the graphs of:
>
> $\quad$ a. $\ y = 3f(x)$ $\quad$ b. $\ y = \frac{1}{2}f(x)$ $\quad$ c. $\ y = f(2x)$ $\quad$ d. $\ y = f(\frac{1}{2}x)$

Activities such as these can be quite useful in helping students understand the connections between a function and its graph, but these connections are *properties of the higher–order function* G. Since G acts on $\mathbf{R}$ to $\mathbf{R}$ functions and produces subsets of $\mathbf{R}^2$, understanding the properties of G requires that students see the $\mathbf{R}$ to $\mathbf{R}$ functions on which G operates (and the subsets of the plane that G produces) as *objects*. In the first activity, the functions under discussion are actually *functions of their coefficients*, and in the second activity, the inputs and outputs to the function being graphed are "pre–processed" and "post–processed," making implicit use of the idea of function composition. So, these activities are really about the functorial or algebraic properties of a representation (G) for certain kinds of functions and certain kinds of transformations on these functions. To be more precise, let's describe the activities in mathematical terms, in order to point up some of the many subtleties inherent in them.

$\quad$ The first activity can be described like this

> Let S be the set of functions $f_{(a,b)}$ defined on $\mathbf{R}$ of the form $x \mapsto ax + b$, and let T be the set of lines in $\mathbf{R}^2$. Given any mapping ϕ on $\mathbf{R}^2$, describe how the line $G\!\left(f_{\phi((a,b))}\right)$ is related to the line $G\!\left(f_{(a,b)}\right)$. In other words, G not only maps elements of S to lines, but it provides a way to induce affine transformations on T through transformations on S, so that G becomes a mapping
>
> $$G\!:\!\left(S, \text{mappings on } \mathbf{R}^2\right) \to \left(T, \text{affine transformations on lines}\right)$$
>
> and it is the functorial properties of *this* correspondence that is studied in these activities. So, the elements of S are considered as functions of their coefficients, and the behavior of G with respect to changes in coefficients is studied. This activity is often extended to polynomial functions (at least to quadratic functions), and Schwartz (1992) outlines activities in which similar properties for the inverse of G can be investigated.

And the second activity becomes:

> Let S be a set of functions on $\mathbf{R}$ (say, polynomial functions or continuous functions). Let $g(x) = ax + b$. Given any $f \in S$, describe $G(f \circ g)$ and $G(g \circ f)$ in terms of $G(f)$. In other words, students are asked to describe changes in the graph of a function when its inputs or outputs are subjected to a linear substitution (that is, when its inputs are *pre–processed* or its outputs are *post–processed*). These activities ask students to investigate special cases of G's behavior with respect to the algebraic operation of composition on S.

$\quad$ Students who come to precalculus with a process notion of function often use the stimulus provided by questions and activities about G to encapsulate $\mathbf{R}$ to $\mathbf{R}$ functions into objects and to develop graphical intuitions that allow them to reason effectively about geometric properties of continuous change. But there are all too few of these students, and we have all seen the effects that graphing activities have when students have not interiorized processes. Because they have nothing to fall back on, students often change the activities, so that G acts on *equations* rather than functions, and then they simply learn the connection between transformations of equations and transformations of graphs. As often happens, when they forget the rules, we hear questions like, "If you subtract 3 from x, does the graph go left or right?"[7] One reason that students change the activity from one where G acts on $\mathbf{R}$ to $\mathbf{R}$ functions to one where G acts on equations is

---

[7] Indeed, a polynomial equation is often seen by students as a convenient coding mechanism for information about a graph. Certain combinations of coefficients produce certain kinds of information about the graph; this information could be simply written down in a table, but for some reason, mathematics teachers like to hide it in an equation.

that most students simply have not developed a process concept of function, and the curriculum provides them with no experiences about function as process. On the other hand, the curriculum *does* provide ample experience with equations and techniques for their manipulation; when all else fails, equations can be used to generate ordered pairs that can be plotted.

This criticism of the **R** to **R** curriculum should *not* be taken as a criticism of visualization in learning about functions. It simply argues that more work on function as process (and object) needs to be in the background of students before they attack the subtleties of the representation G. In fact, since **R** is the underlying set on which functions in this curriculum act, it should be possible to devise environments in which dynamic visualization can be exploited to encourage the development of the process concept for functions. That's exactly the effect of some work by Paul Goldenberg and Phil Lewis (1992). In their environments, functions are visualized by picturing the domain and the range in different windows of a computer screen. As a pointer is moved around the domain, the image moves in the range. For **R** to **R** functions, these two windows simply picture number lines, but the idea works equally well for functions defined on subsets of $\mathbf{R}^{1 \text{ or } 2}$. With **R** to **R** functions, the number lines are arranged one beneath the other in parallel windows. This set–up allows students to investigate the dynamics of the functions *processes*. The environment also allows students to define their *own* functions (as procedures in a computer language), strengthening their view that functions are not the same as equations. Even traditional Cartesian graphs of functions can be introduced in this dynamic format. These *dyna–graphs* are promising tools not only for the introduction of the dynamics of functions, but for more advanced topics like fixed points for functions from **R** to **R** and the properties of the sup metric for the space of continuous functions from a compact subset of **R** to a subset of $\mathbf{R}^2$.

The dyna–graph idea develops and exploits a process concept of function that goes back at least to Euler:

> If some quantities so depend on other quantities that if the latter are changed the former undergo change, then the former quantities are called functions of the latter.

(See Bottazzini, 1986, for example). This notion of (piecewise continuous) *dependence* is at the heart of real and complex analysis, and it conjures up process images that have input–output properties *and* continuity properties that spring from the topology of **R**. Over the centuries, the **R** to **R** curriculum has de–emphasized the process facet of the function idea and has concentrated on the algebraic formalisms of equations. In the classic 1887 Hall and Knight text, *Higher Algebra*, we read:

> Any expression which involves x, and whose value is dependent on x is called a *function of* x.

In 1912, Burnside and Panton (Burnside, 1960) put it this way, "Any mathematical expression involving a quantity is called a *function* of that quantity." Many modern texts have evolved definitions that are close in spirit to Brown (1984):

> A *function* consists of the following:
>
> • A set of real numbers called the *domain* of the function.
>
> • A rule that assigns to each element in the domain exactly one real number.

Anecdote 1: Some effects of the R to R curriculum.

In 1989, seven-high school seniors in my school were working on independent study projects that centered around writing computer programs to do various tasks. One of the projects was to design a graphing program that illustrated the trapezoid rule. Another modeled the Tower of Hanoi puzzle. The other projects were similar; none used the process concept of function in an essential way, and several were based explicitly in the **R** to **R** curriculum. I gave the students a questionnaire containing the following questions:

1. What is a function?

2. What do you think of when you see the word "function"?

The answers to the first two questions show that most of the students thought of a function as a formula or as an equation. Two students described a function as a process, and one described a correspondence between sets (although she incorrectly described single–valuedness).

3. How do you graph a function?

The answers centered around analytic techniques (find the zeros, find the turning points, look for symmetry, etc.)

4. Draw the graph of the equation $y = 2x + 1$.

5. Draw the graph of $f(x) = x^2$.

Everyone was able to draw correct graphs; these graphs don't lend themselves to the analytic bags of tricks, but everyone nicely reverted to plotting points.

6. If $f(x) = x^2 + 1$, what is $f(7)$?

7. This is the graph of $y = g(x)$.

(there was a sketch of a graph here)

Use the graph to estimate the output from g if the input to g is 3. List all the inputs to g that produce an output of 5.5.

Everyone could find $g(3)$ from the graph, but most people found only one pullback of 5.5 (there were two).

8. If h is a function and $h(a) = h(b)$, is $a = b$? Why or why not?

Everyone except the two people who described a function as a process incorrectly answered this question.

9 Describe one function that might have this for a graph:

(there was a sketch of a line here; the axes were labeled z and w)

Only one student used the variables z and w in her description of the function. Two people left it blank, and everyone else used x and y, (two of these gave the wrong function even if you ignore the substitutions $z \mapsto y, \ w \mapsto x$).

10. The length of a rectangle is twice its width. Express the area of the rectangle as a function of its length.

11. In this school, there are about 20 students for every teacher. Express the number of teachers (T) as a function of the number of students (S). Write a Logo function **CONVERT** that takes the number of students as input and outputs the number of teachers.

Only three students were able to correctly answer both the last two questions. At least one student said that the number of teachers is 20 times the number of students.

Questions 3 through 9 are clearly rooted in the **R** to **R** curriculum, and they seemed to set the tone for the students' responses to the first two questions. In particular, most of the students were bound to the

$$y = f(x) = \text{ a polynomial in } x$$

pattern. According to one student (a calculus student at that),

A function is a problem that involves mathematical computation. You can solve a function in a number of ways. When I think of the word function, I think of the many things a function could be: quadratic, polynomial, cubic, etc.

Conclusion 1

Students in the **R** to **R** curriculum do not develop a process concept of function in their coursework.

Conclusion 2

Computer programming, by itself, does not encourage students to see a connection between the already interiorized processes of elementary school and the functions of the **R** to **R** curriculum.

ACTION TO PROCESS: AN EMPIRICAL PROPOSITION BECOMES A RULE

By the time they reach adolescence, most people know that adding 1 to 532 produces 533, and they do not view this as an empirical fact that is open to experimentation. Compare this with the experiments young children perform with blocks or counters when they are first learning about whole numbers. At some point in a child's development, the experiment of "adding 1" becomes unnecessary, and the child can say with certainty that adding 1 to 6 will produce 7.

Wittgenstein (1983) refers to this process by saying that the child has "hardened an empirical proposition into a rule," so that the correctness of the calculation $6+1$ is judged by whether or not the result of the calculation agrees with the (newly formed) rule $6+1=7$. This is very much the same as saying that the child has *interiorized* the process that we describe as the "operation of adding 1" (and that we represent symbolically by $x \mapsto x+1$).

Many operations from algebra are easily interiorized into processes. Consider complex conjugation, for example. High-school students freely describe a manipulation with complex numbers using language like "take the conjugate," and the fact that the complex conjugate of $3+2i$ is $3-2i$ is not seen as an empirical result. Rather, it is seen as the result of applying a *process*.

What activities for students can help them develop a process model for mathematical functions? Here are two that have met with some success.

Anecdote 2: The effect of an alternative curriculum

In 1990, 10 students were working on independent study projects. The students were not all seniors and were taking a variety of courses from geometry through linear algebra and calculus. Before the actual work on their projects, the class worked through a ten week unit in which the primary focus was on experimenting with computer models for functions. Functions were modeled as Logo procedures, and the experiments consisted largely in looking for patterns in the tabulations of the functions. No analytic properties of $\mathbf{R}$ were mentioned, and Cartesian graphs were not used. One activity that the students enjoyed was to take a Logo model for an inductively defined function on $\mathbf{Z}^+$, for example

$$h(n) = \begin{cases} 1 & \text{if } n=1 \\ 3+h(n-1) & \text{otherwise} \end{cases}$$

and to experiment with its tabulation, gradually producing a closed form for the function. Then students were asked to prove that their two functions were equal on $\mathbf{Z}^+$ using mathematical induction (similar experiments were carried out by Sfard (1992). An important feature of the activity is that the closed forms for their functions were not given at the start; students often spent two or more class periods in a search for a closed form for a particular function, and the process involved creating temporary functions, subtracting these guesses from the original function, trying to model this difference, and so on. In short, they had plenty of concrete practice in constructing processes and manipulating them. It is perhaps little surprise that these students had a very different view of functions from the ones expressed by students from the previous year. They answered a questionnaire that had some of the same items as the one mentioned above, but it also carried a quite different tone:

1. What is a function?

2. What do you think of when you see the word "function"?

3. When are two functions equal?

4. Is **SIGMA** a function? Why or why not?

5. If h is a function and $h(a)=h(b)$, is $a=b$? Why or why not?

6. The length of a rectangle is twice its width. Express the area of the rectangle as a function of its length.

7. Here's a problem from a textbook:

    Prove that

$$4 + 4^2 + 4^3 + \cdots + 4^n = \frac{4}{3}\left(4^n - 1\right)$$

Restate the problem in a way that makes use of functions.

8. In this school, there are about 20 students for every teacher. Express the number of teachers (T) as a function of the number of students (S). Write a Logo function **CONVERT** that takes the number of students as input and outputs the number of teachers.

9. For this question, forget about all the mathematics courses you have taken. If f and g are two functions that take numbers as input and produce numbers as output, what would be a reasonable meaning to attach to the expression $f + g$?

10. Carefully describe the general approach you take when you are looking for a closed form for a function.

11. Describe the approach you'd take if you were faced with this problem:

Find a formula for the product of the first n even integers.

For our purposes, the interesting questions are the ones that are also on the previous questionnaire. This time, nine of the 10 students answered the first question by describing a function as a process that takes an input, manipulates it, and produces an output (the inputs and outputs were most often described as "numbers"). Of the six students who took the second question seriously, four actually drew a function machine icon, and the other two gave more detailed descriptions of processes. All but three students knew that a function need not be injective, and six students correctly expressed the area of the rectangle as a function of its length (the others correctly expressed the area as a function of its width). It's not surprising that all the students correctly answered questions 3 and 9, a fact that suggests that these students had begun to encapsulate processes into objects.

After the unit on mathematical induction, two of the students wrote a simple Logo program that graphed **R** to **R** functions by modifying the Logo tool that they had used to tabulate functions over a set of integers (this tool was used by typing a line like

Tab "f 1 10

where f is the name of the function being tabulated and 1 and 10 are the lower and upper bounds for the tabulation). They simply changed the line that printed input–output pairs to a line that *plotted* the pairs on a coordinate system, and they adjusted the increment from 1 to something smaller. They referred to their creation as a "visual tabulator," and, when giving a talk on this tool to the rest of the class, one of the two students said, "I can make the increment small enough to make the graph look like a continuous curve, but don't forget, these are just points obtained from inputs and outputs."

Anecdote 3: Another short example

Five high-school freshmen, taking an introductory programming course in 1990, were modeling functions in Logo and the Function Machine Lab (Feurzeig et al, 1988) (an iconic programming language developed by BBN to help students visualize processes). One of their tasks (on which they worked as a group) was to find as many "different" procedures as possible that tabulated the same (on **Z**$^+$) as a given function.

The students became quite facile in moving between closed form and inductive definitions for the same function (without *any* concern for proof), in summing arithmetic sequences, and so on. When stuck for ideas, they often used equivalent algebraic expressions ($x + x$ for $2x$) in defining their alternate versions, but they always admitted that this kind of thing doesn't "really" give an alternate definition.

One of their problems was to find alternate forms for the function f where $f(x) = (2x+1)(2x-1)$. They produced several, and then looked at the table for f:

$$
\begin{array}{ccc}
1 & \cdots & 3 \\
2 & \cdots & 15 \\
3 & \cdots & 35 \\
4 & \cdots & 63 \\
5 & \cdots & 99 \\
6 & \cdots & 143 \\
& \vdots &
\end{array}
$$

They announced another algorithm that produced this table:

> *To get any output*
> * *add the input to the next input*
> * *add the input to the previous input*
> * *multiply these answers*

So, for example, 99 is calculated as $(5+6) \times (5+4)$. When we tried to write a Logo procedure that modeled this, the students looked with surprise at their alternate form:

```
to f :x
op (:x + (:x + 1)) * (:x + (:x – 1))
end
```

When I claimed that this was just a simple algebraic rearrangement of the original formula defining f, they protested in unison (and quite loudly), "Yeah, but we got there by a different *process*." They spent a good deal of time discussing when two different processes defined the same function, and, although they came close several times, they never quite produced the mathematical definition of function equality.

Conjecture 1

> Activities that focus on finding alternate procedural descriptions can help students think of a mathematical function as a process. Writing descriptions of functions in a high–level programming language encourages students to construct a process model for functions.

It's important not to divorce the second sentence of this conjecture from the first: Simply constructing models for functions on a computer need not lead students to a process concept of function, as can be seen from the behavior of the students described in anecdote 1.[8]

Conjecture 2

> A tabulation for a function is a valuable representation if the goal of the representation is to encourage students to construct mental models of functions as processes. This is especially true if the tabulation develops dynamically on a computer.

A tabulation is an intrinsic visualization for a function. It uses no external information (topological properties of the function's domain or range, for example), and it exists in situations where other forms of visualization are impossible. If the general operation of tabulating a function over a set is provided as a tool in a mathematical programming language, then the very act of asking for a table of values encourages students to think of a function as an object that can be manipulated (in this case, a function is an input to the tabulation tool).

PROCESS TO OBJECT: SEEING RULES AS THINGS

The students described in anecdotes 2 and 3 all talk about functions as processes. Both groups can talk about what a specific function "does" to its input without using specific examples of inputs. Both groups have become quite facile at finding alternate (and often clever) procedural descriptions for a given function. But there is a definite difference between how the students in anecdote 2 think about functions and the way their younger classmates in anecdote 3 think about functions: The anecdote 2 students uniformly say that two

---

[8] More results on the connections between modeling functions in a programming language and the development of a process concept for functions can be found in Breidenbach, et al (1992).

functions are equal if they produce the same output for a given input, while the anecdote 3 students aren't so sure. They sometimes agree with the standard definition, but equally often they claim that the algorithm that leads to a function's outputs contributes in some way to the function's identity.

The difference in attitudes is easy to explain. Both groups are given similar activities in which they must determine alternate descriptions for a function, but the more advanced students must also *prove* that their alternate definitions are correct. The actual *statements* of their problems ("prove that $f = g$ on $\mathbf{Z}^+$") ask them to call two processes equal if they act the same.

Many authors claim that students see the need for function as object when they see the need to manipulate and combine functions. This should happen (but often doesn't) in calculus, when they see the derivative as a way to transform one function into another and when they see theorems like "the derivative of the sum is the sum of the derivatives." It should also happen in linear algebra, where an algebra for linear mappings is constructed in parallel with matrix algebra. It's certainly true that the whole power of the function-as-object idea is that encapsulated processes can be manipulated in ways that produce deep mathematics. But, just as manipulations with vectors require a notion of vector equality, a prerequisite to students' seeing the need for manipulations with functions is that they are able to tell when two functions are equal.

Conjecture 3

> Students who claim that two processes define the same function if the processes produce the same output for a given input have the necessary cognitive development to think about function as object. One way to bring about this development is to ask students to provide arguments which show that different processes tabulate the same.

This notion of identifying processes is not without mathematical precedent; Bourbaki talks about *equivalent* functional relations' determining the *same* function. Think of the set S of processes on, say, $\mathbf{Z}$, realized as the set of computer programs that accept an integer as input and produce an integer as output. The relation $\approx$ is defined on S by $P \approx P'$ if P and P' tabulate the same. Then conjecture three amounts to saying that students are ready to deal with functions as objects when they can work with classes of algorithms modulo $\approx$ in $S/\approx$ rather than with the actual algorithms in S.

Computer environments in which procedural models for functions can easily be defined provide laboratories where students can make processes concrete. Similarly, environments in which functions can be manipulated provide laboratories where students can develop the object concept of function. The next anecdote describes an attempt at constructing such an environment.

Anecdote 4: What do teachers think about functions?

During the summers of 1989 and 1990, thirty high school teachers from Ohio attended a residential six-week program at Kent State to take a course called "Using Computers to Learn Mathematics." This course is part of a much larger endeavor, the Institute for Secondary Mathematics and Computer Science Education, funded by NSF and attended in 1990 by 180 high-school teachers. The CLM course considered several mathematical topics (geometry, combinatorics, and number theory, for example) but the emphasis never strayed very far from the function concept. Although there was considerable interplay between process and object points of view (as well as some applications of computing to the $\mathbf{R}$ to $\mathbf{R}$ curriculum), the first summer's activities were mainly concerned with function as process and the manipulation of processes, and the second summer moved toward modeling these manipulations as higher–order functions.

Most participants came to IFSMACSE thoroughly immersed in the $\mathbf{R}$ to $\mathbf{R}$ curriculum. When asked to define functions, the overwhelming majority of participants described the "vertical line test" for $\mathbf{R}$ to $\mathbf{R}$ functions. Others talked about sets of ordered pairs or about correspondences between sets. However, informal conversations with participants showed that most (say, 25 of the 30) also viewed functions as processes (although many used the context of an algebraic formula relating x and y); they felt compelled to give more formal definitions to their students, because that's the way most textbooks do it. The process description of function was a heuristic that many participants used here and there in their classes.

During the first summer, participants modeled functions as Logo procedures and as Function Machines. Some of their activities simply asked them to model traditional functions from precalculus and to use tabulations rather than graphs as a medium for experimentation. Other activities (similar to ones given to the high school students described in anecdote 2) asked them to find closed forms for inductively-defined functions and to prove that the resulting pairs of functions were equal. Examples of both types of activities are given in Appendix A.

Participants certainly saw the point of these activities; unlike the high school students, many were not constructing the process concept of function for the first time. Several teachers said that they had been using the function machine metaphor for years when explaining **R** to **R** functions. Still, it was clear that the participants considered the process viewpoint secondary to the viewpoint put forth in the **R** to **R** curriculum, and that their thoughts about general processes were heavily influenced by the specific processes used in precalculus. As an example of the degree to which the **R** to **R** curriculum exerts its hold, consider the remarks of one participant:

> … The definition I recall is that for each element of the domain, there is one and only one element of the range paired with it. I always related the concept of functions to ordered pairs (x, y) where x and y are elements of the real number system. One could picture a function as an xy–graph and perform the vertical line test to illustrate that no element of the domain is ever repeated.

> … I no longer see functions as simple ordered pairs of numbers. Functions are machines that input "things." I see the need to be very flexible and view functions as generators of sets of "things" that still somewhat conform to the old rules; however the domain and the range (image now) can be anything, not just simple reals.…

The participants then worked through problem sets that asked them to manipulate processes. For example, given a function f defined on **Z**, they constructed a function g defined by

$$g(n) = \sum_{k=1}^{n} f(k)$$

One of the goals of the activities was for participants to begin to interiorize these higher–order manipulations into processes. In the example just mentioned, g is really a function of f, and this accumulation operator that transforms f to g is a function defined on *functions*.

These activities had mixed success. One of the difficulties is that we were pushing the media (essentially Logo) to the limits of its utility. Some participants *did* manage to model higher–order functions in full generality (they essentially constructed a **lambda** in Logo), but the task required a high–powered and technical excursion into list processing.

Another difficulty was that, for many participants, the computer experiments with manipulations of processes simply didn't inspire the image of one process being transformed into another. Most were perfectly willing to use one function as a *parameter* in another function, but it wasn't seen as the primary input; *that* always had to be a number. For example, one of the activities asked participants to approximate the definite integral of a continuous function. This approximation A was uniformly seen as a function of the upper limit of integration. Participants built models of A that captured the idea

$$A(x) \approx \int_{0}^{x} f(t)\, dt$$

After several exercises where they had to edit the code for A simply to change the name of the function being integrated, they gave the function the status of an input and changed their model to

$$A(f, x) \approx \int_{0}^{x} f(t)\, dt$$

The change here was simply for convenience; the important input to A was x, not f. Indeed, when it was suggested that A can be viewed as a function of f, many participants saw absolutely no difference between the idea expressed in

$$A(f)(x) \approx \int_{0}^{x} f(t)\, dt$$

and the idea expressed by their two–input model for A.

One of the activities contained the following problem (which is quite difficult to do in Logo):

> The derivative of a function f is a function that can be approximated by the following formula

$$\mathbf{APPROX.DERIV}(f)(x) = \frac{f(x + .0001) - f(x)}{.0001}$$

Write a Logo model for **APPROX.DERIV** that takes *only one* input.

One group of participants worked on this problem for an entire afternoon. That evening, one of them asked me, "Shouldn't **APPROX.DERIV** take a *function* as input as well as a number?"

The function activities for the second summer were explicitly directed at developing the concept of function as object. Most of these activities were aimed at creating disequilibrating situations for the participants, so that they would re–examine their ideas about functions.[9] The epistemological perspective driving the approach implies that someone will encapsulate processes by acting on them and by transforming them. This implies that the object concept of function is tied to the notion of higher–order functions that take functions as input and/or return functions as output. Two examples of these higher–order functions are shift operators (things like $f \mapsto (x \mapsto f(x+1))$) and function addition. In other words, one way to see functions as objects is to see them as data. In order to model these higher–order functions on a computer, the participants used ISETL.

ISETL is an outgrowth of SETL, a language developed by Schwartz et al (1980) in which the mathematical notion of set is a primary data type. Ed Dubinsky thought about using the language in mathematics classrooms, and he enlisted the help of Gary Levin, a computer scientist at Clarkson, who wrote an interactive and interpreted version of the language in C. ISETL has evolved through several versions, but, from the start, the development of the language has been strongly influenced by the concurrent research of Dubinsky and others on how people learn mathematics. The language was first introduced to students in a discrete mathematics course at Berkeley early in 1985. An accessible reference to ISETL is Dauterman (1993). In ISETL, sets and functions are first-class objects, and the syntax of the language is remarkably close to standard mathematical notation. Once one is proficient in the low–level system details, using ISETL feels not so much like programming as it feels like writing mathematical statements. A brief description of how one models sets and functions in ISETL is given in Appendix B.

The IFSMACSE activities began by having participants model functions as ISETL **funcs** (procedures that output) and as sets of ordered pairs. A new use was made for a set of ordered pairs for a **func**: It is straightforward to write an ISETL procedure that takes any finite set and tabulates it (simply lists its elements vertically on the computer screen); when the input to this tabulation tool is a set of ordered pairs for a **func**, the resulting display gives a tabulation for the **func**. For this reason, we called the set of ordered pairs for a **func** the *table* for the **func**. Teachers practiced changing from one representation to another, until they were able to write ISETL functions **FUNCtoTABLE** and **TABLEtoFUNC** that changed representations programatically. Tables were also used as inputs to predicates that tested for functionality and for equality of functions.

The participants worked through problems where, for example, they described the difference between the function **shift** defined by:

$$\textbf{shift}(f, x) = f(x+1)$$

and **shift1**:

$$\textbf{shift1}(f) = (x \mapsto f(x+1))$$

One problem in particular seemed to be quite disequilibrating to the class, and it marked a turning point for many of the people who went on to demonstrate that they had a solid notion of encapsulated processes as objects. The instructions were:

> Construct an ISETL table for each function on the given set. Do something with each table (like name it, **TAB** it, and test it for "function–ness" with your **functionp** predicate). Look for functions that are equal. Look for functions that are almost equal. Look for functions that are not equal. Look for other stuff.

There followed a list of a dozen functions, all having subsets of $\mathbf{R}^{1\text{ or }2}$ as domain and range ($y \mapsto (y, y^2)$ on $\{-3, -2.9, -2.8, \ldots, 1\}$, for example). The last function to be investigated was

$$x \mapsto (a \mapsto a^2 + x) \text{ on } \{-1, \ldots, 5\}$$

Everyone immediately wrote down the ISETL code for this function, because it is a direct translation of the mathematical notation; the ISETL model is:

---

[9] These activities are clearly inappropriate for many high school students, and a prerequisite to developing activities that require high-school students to work with higher-order processes will be to develop meaningful situations in which such processes are *useful*. Many such situations already exist in the **R** to **R** curriculum, and the infusion of topics from "discrete" mathematics into the high-school curriculum will provide even more opportunities for students to work with functions as objects.

$$| \; x \to |a \to a*a + x| \; |;$$

The resulting **func** was disquieting to the entire group. In spite of the fact that the translation to ISETL is immediate, people traded descriptions of the function that showed fundamental confusion. For example:

- Some participants claimed that the function produced a number. They described the function as its output: $a \mapsto a^2 + x$.

- In an attempt to describe the output as a function, some people said things like "you'll get a parabola". Some of *these* people then changed their minds and claimed that the output is a "line" because x isn't squared.

- This function assigns numbers to functions. This means that it can be evaluated at numbers and tabulated over sets of numbers. An ISETL tabulation for the function produces something like:

    1 ... !func(39)!
    2 ... !func(40)!
    3 ... !func(41)!
    4 ... !func(42)!
    5 ... !func(43)!
    6 ... !func(44)!
    7 ... !func(45)!
    8 ... !func(46)!

    The right hand column is ISETL's mechanism for reporting a functional object. Participants had seen this mechanism before, but about half of them took this tabulation as an error message. Several expressed the view that the **!func(??)!** output was acceptable if the input was a function, but *not* if the input was a number. One participant reported that she worked on this problem for some time, changing and refining her ISETL model for the function (which was perfectly accurate in the first place), until she got the tabulation to produce number outputs, "like it's supposed to."

This example was *not* the first time the class had considered functions that output other functions. Why did it cause such a stir?

- Previous examples had discussed transformations on functions (like **shift**) that accept functions and produce functions. This example produced a function from a number. The fact that this variation was so disequilibrating points to the possibility that participants had not been constructing images for "functions that produce functions" but rather "processes that change functions."

- The fact that this function could be tabulated led people to think that it should behave like all the other functions they had tabulated; *these* had always produced elements of $\mathbf{R}^n$.

- When something is confusing, we try to accommodate it into our existing cognitive structures, and, when new structures are not quite established, we relate the disquieting phenomena to old and established notions. The form of the output of the function's output, $a^2 + x$, was close enough to the $\mathbf{R}$ to $\mathbf{R}$ curriculum's classic examples (the *parabola* and the *line*) and far enough from newer ideas of higher–order functions that participants tried to accommodate the example into the older, more established notions of function. The $\mathbf{R}$ to $\mathbf{R}$ curriculum's rule about letters' at the beginning of the alphabet representing constants and letters' at the end representing variables combined with the rule that "squaring" points to a quadratic function, caused participants to oscillate between viewing " $a^2 + x$ " as a "line" and a "parabola". (Watching them discuss this was quite reminiscent of watching someone studying a non–perspective two-dimensional representation of a wire-frame cube, where the perception of what is the "front" of the cube keeps changing.)

- The people who were most confused by the function were the people who insisted on relating it to **R** to **R** functions. On the other hand, all the people who found an explanation of the function that led to re–equilibration did so by talking about functions as processes. One participant described his initial confusion by saying that he thought that the output of the function was a sausage when, in fact, it was a meat–grinder. This analogy became a staple of the class discussions, and it helped several people move toward encapsulating processes.

- The "breakthroughs" in participants' attempts to understand the function sprung from their attempts to explain the strange behavior of its ISETL model (its tabulation, for example). In other words, people used their computer implementation of the function as an important part of the re–equilibration process.

Conclusion 3

Working in a computer medium that uses a system of notation which allows for direct translation of mathematical statements does not by itself insure that students will build the mental constructs necessary to understand the mathematical statements.

In other words, the fact that the ISETL code for this function was almost identical to the mathematical notation had no effect on participants' initial understanding of the function; in fact, most people simply typed in the ISETL code and tried to understand the function *from* the ISETL model. One benefit, then, in having a mathematical programming language is that mathematical abstractions can be implemented on a computer with little overhead, and this gives the students a concrete object on which they can experiment and reflect:

Conclusion 4

Working in such a medium *does* provide an environment in which discussions and experiments that lead to an understanding of the mathematical statements can be supported.

The final activities in this part of the course centered around building ISETL models for arithmetic operations with **R** to **R** functions, models for composition and iteration of arbitrary functions, and models for more general higher–order functions. Some examples of these later activities are given in Appendix C. Some members of the class didn't work on these activities at all. Of the other participants (approximately 25), about half carried on discussions that showed that they could move between interpretations of functions as processes and as objects. One group of four participants produced a final project that used higher–order functions in an essential way.

SOME MORE CONJECTURES AND CONCLUSIONS

We often hear that the mathematics of formal education is quite different from the mathematics done by mathematicians. Attempts at closing this gap often center on the differences between the way students work and the way researchers work, and proposals for reform in mathematics-education concentrate on changing the ways students acquire mathematical knowledge. The NCTM *Standards* calls for experimentation, for conjecturing, and for long-term mathematical projects carried out by students working cooperatively in groups. The *COMET* report asks pre–service teachers to carry out significant research projects as a formal part of their training.

As important as these attempts to make mathematics an experimental science are, there is another (equally important) gap between the mathematics of the mathematician and the mathematics of the non–specialist: mathematicians use, as a routine part of their work, a form of encapsulation that non–mathematicians find quite foreign. This peculiarly mathematical activity involves a bending of ordinary language in a Wittgenstein–like language game in which distinctions between objects and methods are dissolved. People outside mathematics find it odd that mathematical researchers find this activity interesting, but, over and over again, this stretching of the language of mathematics has brought about important advances. Indeed, the crisis in the foundations of mathematics came about from attempts at encapsulation (the "set of all sets," for example) that force us into contradictions.

Finding environments in which students develop a knack for the encapsulation of mathematical processes is a different task from the one taken up by the current grade 10–15 curriculum in which the primary goal is to expose students to some important results and techniques from classical analysis. Students can, in fact, experience the dizzying effects of mathematical encapsulation with very *little* background, especially in the case of functions. Consider the following example of how functions can be interchanged with objects, keeping in mind the impression that this example would have on a student:

> Let T^S denote the set of functions from S to T. Then we can consider elements of S as elements of $T^{(T^S)}$ by defining $s(f)$ as $f(s)$.

There are no high–powered mathematical constructions here; the example uses only the notions of sets and mappings. But the statement is difficult to understand, and part of the reason for that is in the definition of the duality $s(f) = f(s)$, where s and f are used as *both* process and object.

What kinds of activities would be effective in bringing students to the point where statements like this makes sense? The anecdotes in this paper suggest some directions for the case of functions:

- Students need *both* a process and object concept of function in order to work with modern mathematics. Specific activities that involve finding several computer implementations for the same function help students build a process concept of functions. Building an object concept is much harder, but the construction of computer models for higher–order functions helps some people make the necessary constructions. A very perceptive participant at IFSMACSE put it this way:

  My view of functions was limited to processes before IFSMACSE. I never dealt with funcs beyond one unit of composition ….

  "… one unit of composition" refers to the fact that certain operations on functions, like composition, can be performed both on the functions at hand and on the higher–order functions that manipulate them (think of iterated differentiation, for example). More thought needs to be given to the genetic decomposition of the function as object concept, so that activities can be developed that will be effective with a wide audience of students.

- The **R** to **R** curriculum, with its total concentration on real valued functions of a real variable and its implicit identification of a function with its graph, does nothing to encourage students to develop a process concept of function. This seems to be a result of the emphases in the curriculum rather than anything in the nature of the topics. In addition to the dyna–graph idea, having students construct the function

  $$f \mapsto \text{the graph of } f$$

  by representing it as the composite of two more concrete functions

  $$f \mapsto \text{the table for } f \mapsto \text{the graph of the table for } f$$

  might reinforce the notion that a Cartesian graph is just one method for visualizing a process. The use of environments like ISETL to construct each piece of the composite completely removes the tedium involved in constructing tables and graphs by hand, and it allows students to concentrate on the connection between an **R** to **R** function and its graph. A graphing environment that simply plots sets of ordered pairs coupled with an easy mechanism for generating sets of ordered pairs from a process might be a valuable tool in this endeavor.

Sfard (1992) raises the possibility that "… there may be students for whom the structural conception will remain practically out of reach whatever the teaching method." Perhaps. But what was once considered the frontier of mathematical research is now part of the working vocabulary of school children. It's quite likely that the instructional environments evolving from the current interest in the function concept will eventually make the mathematician's view of function accessible to most students of mathematics.

APPENDIX A:
SOME IFSMACSE ACTIVITIES FROM SUMMER 1

This first activity was designed to get participants to use tabulations rather than graphs as a medium for experimentation:

> Write Logo versions of the following mathematical functions. Tabulate each function between 1 and 15. Which pairs have identical tabulations? Why? Do any functions have almost the same tabulations? Can you explain why?

(a) $F(X) = X^3 + 3X - 2$

(b) $G(X) = 5X + 7$

(c) $S(X) = F(X) + G(X)$

(d) $V(X) = \begin{cases} F(X) & \text{if } X > 9 \\ S(X) & \text{otherwise} \end{cases}$

(e) $H(X) = F(X+.001)$

(f) $W(X) = F(G(X))$

(g) $D(X) = \dfrac{F(X+.001) - F(X)}{.001}$

(h) $J(X) = H(X-.001)$

(i) $M(X) = 3X^2 + 3$

The next activity asks people to find alternate descriptions for functions and to prove that their alternate descriptions are correct.

> Give a closed form for each function (you can arrive at the closed form by a tabulation or by other methods). Then prove that your closed form is equal to the given function on $\mathbf{Z}^+$ by the principle of mathematical induction:

a)
```
TO F :N
   IF :N = 1 [OP 3]
   OP (F :N - 1) + 5
   END
```

(b)
```
TO G :X
   IF :X = 1 [OP 4]
   OP (G :X - 1) + 3 * :X
   END
```

(c)
```
TO H :X
   IF :X = 1 [OP 2]
   OP (H :X - 1) + 2 * :X - 1
   END
```

(d)
```
TO J :X
   IF :X = 1 [OP 1]
   OP (J :X - 1) * 2
   END
```

(e)
```
TO K :X
   IF :X = 1 [OP 1]
```

```
        OP (K :X - 1) + (POWER 4 :X)
        END

(f)  TO L :X
        IF :X = 1 [OP 4]
        OP (L :X - 1) +(POWER 5 :X)
        END
```

APPENDIX B:
ISETL, SETS AND FUNCTIONS

The power of ISETL is that it allows students to express the primary mathematical notions of *set* and *function* in almost complete generality and with very little overhead. A set is denoted very much the way it is in mathematics; the ISETL expression

$$\{ \text{x*x} : \text{x in } \{ 1 .. 5 \} \};$$

produces an *output* (rather than a side effect) that is the ISETL representation for the set of squares of the integers between 1 and 5. The actual value that the expression returns might be printed on the screen as

$$\{ 4,9,1,16,25 \}$$

because (just as in mathematics) the order in which the elements of a finite set are generated has no effect on the value of the set.

Functions can be modeled as procedural objects. For example, the function that cubes its input can be represented in ISETL as

```
f := func(x);
return x**3;
end;
```

Just as in mathematics, the notation **f(5)** produces 125. The assignment of the name **f** to this function is purely for convenience, 125 will also be returned if one types

```
func(x); return x**3; end (5);
```

The definition of **f** can also be accomplished by the suggestive shorthand:

$$\text{f} := |\ x \rightarrow \text{x**3}\ |;$$

and |x –> x**3|(5) produces 125.

Functions can also be represented as sets of ordered pairs. If the set

$$\{ [\text{x,x*x} + 1] : \text{x in } \{ 1 .. 5 \} \};$$

is named **g**, then **g(2)** produces 5. Converting from one representation of a function to another is simple, and the conversion process can be captured and modeled as an ISETL function.

As an example, consider the problem of mapping a function over a set. This typical recursive algorithm can be modeled in ISETL, but one can also describe a process that makes the element-by-element construction implicit:

```
map := func(f,set);
return { f(x) : x in set };
end;
```

This says, "To map f over S, return the set of all $f(x)$ as x runs over S." Assuming that the prerequisite cognitive structures (elementary set theory, for example) are in place, students find the simple elegance of this faithful model of the underlying mathematics quite compelling.

Higher–order functions can be modeled in ISETL using the same mechanism one uses to construct any function. As an example, one way to model function composition is:

```
compose := func(f,g);
  return func(x);
    return f(g(x));
  end;
end;
```

This can also be written as:

```
compose := | f,g –> |x –> f(g(x))| |;
```

APPENDIX C:
SOME IFSMACSE ACTIVITIES FROM SUMMER 2

These problems came after participants had built ISETL models for function algebra with **R** to **R** functions.

38 Suppose you have a function f. The *nth iterate* of f, written f^n, is the function

$$f^n = \underbrace{f \circ f \circ f \circ \cdots \circ f}_{n \text{ times}}$$

For example, if $f(x) = 3x+1$, then f^3 is the function $f \circ f \circ f$, so that, if a is a number,

$$f^3(a) = f \circ f \circ f(a)$$
$$= f\big(f(f(a))\big)$$
$$= f\big(f(3a+1)\big)$$
$$= f(3(3a+1)+1)$$
$$= 3(3(3a+1)+1)+1$$

which is a thing that should be simplified. So, write an ISETL higher–order func **ITER** that takes an integer n and a function g. **ITER**(n,g) returns g^n.

39 Consider the following ISETL **func**

```
r := func(n);
return ITER(n,cos);
end;
```

Describe $r(n)$ in words. Let $f = r(30)$. Describe f.

40 Notice that you designed **ITER** to work on functions, but it also works on higher–order functions, like the derivative. If $f(x) = x^4 + 5x + 1$, what is the fifth derivative of f? Use your ISETL models **ITER**, **deriv**, and a model for f to check this out. Don't forget, **deriv** is only an approximate derivative.

41 Build an ISETL model for the higher–order function ψ where

$$\psi(f)(x) = xf(x) + (x+1)f(x+1)$$

Let **one** be the function that always outputs 1, and consider the functions s defined by

$$s(n) = \begin{cases} \textbf{one} & \text{if } n = 0 \\ \psi^n(\textbf{one}) & \text{otherwise} \end{cases}$$

Model s in ISETL. Each function $s(n)$ is a polynomial, and two of my students once spent a wonderful semester exploring the properties of these polynomials. Write out a few of the $s(n)$ by hand. See anything? The constant terms of these polynomials (that is, the $s(n)(0)$) are connected with an interesting combinatorial problem that comes up in the real world ...

PART V

REFERENCES

An Exploration of the Nature and Quality of Undergraduate Education in Science, Mathematics, and Engineering, A Report of the National Advisory Group of Sigma XI, the Scientific Research Society, Wingspread, 1989.

Adams, N. A. & W. R. Holcomb. (1986). Analysis of the relationship between anxiety about mathematics and performance. *Psychological Reports,* **59,** 943–948.

Adcock, A., J. R. Leitzel & B. K. Waits. (1981). University mathematics placement testing for high school juniors. *American Mathematical Monthly,* **88,** 55–59.

Adda, J. (1982). Difficulties with mathematical symbolism: Synonomy and homonomy, *Visible Language,* **16,** 205–214.

Aiken Jr., L. R. (1976). Update on attitudes and other affective variables in learning mathematics. *Review of Educational Research,* **46,** 293–311.

American Association for the Advancement of Science (1984). *Equity and excellence: Compatible goals. An assessment of programs that facilitate increased access and achievement of females and minorities in K–12 mathematics and science education.* Washington, DC: author.

Amit, M. & N. Movshovitz–Hadar. (1991). Applications of R–Rules as exhibited in calculus problem solving. In Furinghetti, F. (Ed.), *Proceedings of the Fifteenth Conference of the International Group for the Psychology of Mathematics Education,* **1,** (pp. 57–64). Genova, Italy.

Amit, M. & S. Vinner (1990). Some misconceptions in calculus: Anecdotes or the tip of an iceberg? *Proceedings of the 14th International Conference for the Psychology of Mathematics Education,* **1,** (pp. 3–10). Oaxtepec, Mexico.: CINVESTAV.

Anderson, J. (1989). Sex–related differences on objective tests among undergraduates. *Educational Studies in Mathematics,* **20,** 165–177.

A profile of 1987 recipients of doctorates. (March, 1989). *Chronicle of Higher Education,* **35,** (27), A13.

Arons, A. B. (1977). Development of the capacity for abstract logical reasoning. *Journal of College Science Teaching,* **14,** 248–249.

Artigue, M. (1986). The notion of differential for undergraduate students in sciences, *Proceedings of the Tenth International Conference for the Psychology of Mathematics Education* (pp. 229–234). London: University of London Institute of Education.

Asera, R. (1988). The mathematics workshop: A description. Unpublished paper.

Astin, A. (1968). *The college environment.* Washington, DC: American Council on Education.

Atwater, M. M. & R. D. Simpson. (1984). Cognitive and affective variables affecting black freshmen in science and engineering at a predominantly white university. *School Science and Mathematics,* **84,** 100–112.

Austin, J. D. (1980). When to allow student questions on homework. *Journal for Research in Mathematics Education,* **11,** 71–75.

Ayers, T., G. Davis, E. Dubinsky & P. Lewin (1988). Computer experiences in learning composition of functions. *Journal for Research in Mathematics Education,* **19,** 246–259.

Bachelard, G. (1983). *La formation de l'esprit scientific* (12th e.). Paris: Librarie philsophique J. Vrin. (Original work published 1938).

Bakar, M. & D. Tall (1991). Students' mental prototypes of functions and graphs. In Furinghetti, F. (Ed.), *Proceedings of the Fifteenth Conference of the International Group for the Psychology of Mathematics Education,* **1,** (pp. 104–111). Genova, Italy.

Ball, D. L. (1990). Prospective elementary and secondary teachers' understanding of division. *Journal for Research in Mathematics Education,* **21,** 132–144.

Bander, R. S., R. K.Russell & K. P. Zamostny (1982). A comparison of cue–controlled relaxation and study skills counseling in the treatment of mathematics anxiety. *Journal of Educational Psychology,* **74,** 96–103.

Barrett, L. K. & W. Browder (1989). Reflections on the calculus initiative. *UME Trends,* **1,** (4), 8.

Barszczewski, M. (1986). *Problem–solving approaches and changes in approach used by community college students.* (Doctoral dissertation, New York University, 1985). Dissertation Abstracts International, **46,** 2219A.

Bassok, M. (1990). Transfer of domain–specific problem–solving procedures. *Journal of Experimental Psychology: Learning, memory and cognition*, **16**, 552–533.

Bauldry, W. C. (1990). The pilot "Calculus with Computers" course at Appalachian State University. In Demana, F. & Harvey, J. (Eds.), *Proceedings of the Conference on Technology in Collegiate Mathematics* (pp. 100–103). Menlo Park, CA: Addison–Wesley Publishing Company.

Baxter, N. (1991). *Introductory Computer Science*, Manuscript.

Baxter, N and E. Dubinsky (in press). A new approach to teaching abstract mathematical concepts. *Journal of Mathematical Behavior*.

Baxter, N., E. Dubinsky & G. Levin. (1989). *Learning Discrete Mathematics with ISETL*, NY: Springer–Verlag.

Baylis, J. (1983). Proof — the essence of mathematics, part 1. *International Journal of Mathematics Education and Science Technology*, **14**, 409–414.

Becker, J. R. (1984). The pursuit of graduate education in mathematics: Factors that influence women and men. *Journal of Educational Equity and Leadership*, **4**, 39–53.

Becker, J. R. (1986). Mathematics attitudes of elementary education majors. *Arithmetic Teacher*, **33**, 50–51.

Becker, J. R. (1989). The pursuit of graduate education in the mathematical sciences. In I. Ravina, & Y. Rom (Eds.). *GASAT 5: Contributions to the Fifth International Conference* (pp. 78–86). Haifa, Isr.: Technion.

Becker, J. R. & B. J. Pence. (April 1990). *The teaching and learning of college mathematics: Current status and future directions*. Unpublished manuscript. California State University, Institute for Teaching and Learning.

Becker, J. R. & B.J. Pence. (1990). Linkages between teacher education and classroom practices. In J. A. Dossey, A. E. Dossey & M. Parmantie (Eds.). *Preservice teacher education: The papers of Action Group 6 from ICME 6*, Budapest, Hungary (pp. 201–206). Normal, IL: Illinois State University.

Beckmann, C. E. (1990). Effect of computer graphic use on student understanding of calculus concepts. In F. Demana & J. G. Harvey (Eds.), *Proceedings of the Conference on Technology in Collegiate Mathematics* (pp. 153–158). Menlo Park, CA: Addison–Wesley Publishing Company.

Belenky, M. F., B. M. Clinchy, N. R. Goldberger & J. M. Tarule. (1986). *Women's ways of knowing: The development of self, voice, and mind*. NY: Basic Books.

Berenson, S. B. & L. V. Stiff. (1990/91). Uses of instructional technologies: First year report on change at a university. *Journal of Computers in Math and Science Teaching*, **10**, 11–20.

Boli, J., M. L. Allen & A. Payne. (1985). High–ability women and men in undergraduate mathematics and chemistry courses. *American Educational Research Journal*, **22**, 605–626.

Bonsangue, M. (1992). *The effects of calculus workshop groups on minority achievement and persistence in mathematics, science, and engineering*. Unpublished doctoral dissertation, Claremont, CA.

Borko, H. & C. Livingston. (1989). Cognition and improvisation: Differences in mathematics instruction by expert and novice teachers. *American Educational Research Journal*, **26**, 473–498.

Bottazzini, U. (1986). *The higher calculus: A history of real and complex analysis from Euler to Weierstrass*, Springer Verlag,: NY.

Bradbury, R. (April, 1982). Presentation made at University of San Diego, CA.

Brechting, M. C. & C. R. Hirsch. (1977). The effects of small group–discovery learning on student achievement and attitudes in calculus. *MATYC Journal*, **11**, 77–82.

Breidenbach, D., E. Dubinsky, J. Hawks & D. Nichols. (1992). Development of the process conception of function. *Educational Studies in Mathematics*, **23**, 247-285.

Brown, R. G. & D. P. Robbins. (1984). *Advanced Mathematics — A precalculus course*. Hopewell, N.J: Houghton–Mifflin.

Brown, S.I. (1971). Rationality, irrationality and surprise. *Mathematics Teaching*, **55**, pp. 13–19.

Brown, S. V. (1987)). *Minorities in the graduate education pipeline*. Princeton, NJ: Educational Testing Service.

Brown, S. I. & M. I. Walter. (1982). *The Art of problem posing*. Philadelphia.: The Franklin Institute Press.

Browning, C. (1990). The computer/calculator precalculus (C^2PC) project and levels of graphical understanding. In F. Demana & J.G. Harvey (Eds.), *Proceedings of the Conference Technology in Collegiate Mathematics* (pp. 114-116). Menlo Park, CA: Addison-Wesley Publishing Co.

Bruffee, K.A. (1987). The art of collaborative learning. *Change*, **19**, (2), 42–47.

Buerk, D. (1982). An experience with some able women who avoid mathematics. *For the Learning of Mathematics*, **3**, 2.

Buerk, D. (1985). The voices of women making meaning in mathematics. *Journal of Education*, **167**, 59-70.

Buerk, D. (December, 1988). Mathematical metaphors from advanced placement student. *Humanistic Mathematics Network Newsletter*, **3**.

Bull, E. K. (1983). *Problem representations and solution procedures used in solving algebra word problems.* (Doctoral dissertation, University of Colorado at Boulder, 1982). Dissertation Abstracts International, **43**, 2279A.

Burger, W. F. (1971). *Effects of varying availability of remediation on ability to construct proofs in a simple mathematical system presented via computer assisted instruction.* (Doctoral dissertation, The Ohio State University, 1974). Dissertation Abstracts International, **35**, 5146A.

Burnside, W. S. & A. W. Panton. (1960). *The theory of equations.* Dover, NY.

Burton, L. (1984). Mathematical thinking: The struggle for meaning. *Journal for Research in Mathematical Education*, **15**, 1, 35–49.

Buxton, L. (1981). *Do you panic about maths? Coping with maths anxiety.* London: Heinemann Educational Books.

Carlson, P. R. (1972). *An investigation of the effects of instruction in logic on pupils' success in proving theorems in mathematics.* (Doctoral dissertation, University of Minnesota, 1971). Dissertation Abstracts International, **31**, 3148A.

Caruso, G. E. (1966). *A comparison of two methods of teaching the mathematical theory of groups, rings, and fields to college freshmen.* (Doctoral dissertation, New York University).

Casserly, P. L. (1979). Helping Able Young Women Take Math and Science Seriously in School. Reprinted with revisions, from Colangelo Zaffrann: *New voices in counseling the gifted*, Kendall, Hunt Publishing Company. NY: The College Board..

Chang, P. (1977). Small group instruction: A study in remedial mathematics. *MATYC Journal*, **11**, 72–76.

Charles, R. I. & E. A. Silver (Eds.) (1989). *Research agenda for mathematics education: The teaching and assessing of mathematical problem solving.* Reston, VA: National Council of Teachers of Mathematics.

Claxton, C. S. & Murrell, P.H. (1987). *Learning styles: Implications for improving educational practices. ASHE-ERIC Higher Education Report 1987*, No. 4. Washington, D.C.: The George Washington University.

Clement, J. & C. Konold. (November, 1989). Fostering basic problem–solving skills in mathematics. *For the Learning of Mathematics.*

Clement, J., J. Lochhead & G. Monk (1981). Translation difficulties in learning mathematics. *American Mathematical Monthly*, **88**, 286–290.

Clute, P. S. (1984). Mathematics anxiety, instructional method, and achievement in a survey course in college mathematics. *Journal for Research in Mathematics Education*, **15**, 50–58.

Cobb, P. (June, 1989). Experiential, cognitive, and anthropological perspectives in mathematics education. *For the Learning of Mathematics.*

Cobb, P., T. Wood, E. Yackel, J. Nicholls, G. Wheatley, B. Trigatti & M. Perlwitz. (1991). Assessment of a problem–centered second–grade mathematics project, *Journal for Research in Mathematics Education*, **22** (1), 3–29.

Cocking, R. R. & S. Chipman. (1988). Conceptual issues related to mathematics achievement of language minority children. In R.R. Cocking & J.P. Mestre (Eds.), *Linguistic and cultural influences on learning mathematics* (pp. 17–46). Hillsdale, NJ: Lawrence Erlbaum Associates.

Cohen, D. W. (1982). A modified Moore method for teaching undergraduate mathematics. *The American Mathematical Monthly*, **89**, 473–490.

Cole, M. & P. Griffin. (1987). *Contextual factors in education: Improving science and mathematics education for minorities and women.* Madison, WI: Wisconsin Center for Education Research.

Collis, K. F. (1974). *Cognitive development and mathematics learning*. Paper presented at the Psychology of Mathematics Workshop, Centre for Science Education, Chelsea College, London.

Confrey, J. (1980). *Conceptual change, number concepts and the introduction of calculus,* Unpublished doctoral dissertation, Cornell University, Ithaca, NY.

Confrey, J. (1990). What constructivism implies for teaching. In R.B.Davis, C.A.Maher, & N. Noddings, (Eds.). *Constructivist views on the teaching and learning of mathematics.* (Journal for Research in Mathematics Education Monograph No. 4). Reston, VA: National Council of Teachers of Mathematics.

Cooney, T. J. (1985). A beginning teacher's view of problem solving. *Journal for Research in Mathematics Education*, **16**, 324–336.

Cope, C. L. (1988). Math anxiety and math avoidance in college freshmen. *Focus on Learning Problems in Mathematics*, **10**, 1–13.

Cope, D. E. & A. J. Murphy. (1981). The value of strategies in problem solving. *Journal of Psychology*, **107**, 11–16.

Copes, L. (1982). The Perry Development Scheme: A metaphor for learning and teaching mathematics. *For the Learning of Mathematics*, **3**, 1: 38–44.

Copes, L. (in press) Mathematical orchards and the perry development scheme. *Humanistic Mathematics Network Journal*.

Counting on You: Supporting standards for mathematics teaching. Mathematical Sciences Education Board, 1990.

Cuoco, A. (in press). An action to process approach to solving algebra word problems, To appear in *Interactive Tutoring Environments*.

Cupillari, A. (1989). *The nuts and bolts of proof.* Belmont, CA: Wadsworth.

Dambrot, F. H., M. A.Watkins–Malek, S. M.Silling, R. S. Marshall & J. A. Garver. (1985). Correlates of sex differences in attitudes toward and involvement with computers. *Journal of Vocational Behavior*, **27**, 71–86.

J. Dauterman. (1993) *Using ISETL — a language for learning mathematics.* West Publishing, St. Paul, MN.

Davidson, N. (Ed.) (1990a). *Cooperative learning in mathematics: A handbook for teachers.* Menlo Park, CA: Addison–Wesley Publishing Co.

Davidson, N. (1990b). The small–group discovery method in secondary– and college–level mathematics. In N. Davidson (Ed.), *Cooperative learning in mathematics: A handbook for teachers* (pp. 335–361). Menlo Park, CA: Addison–Wesley Publishing Co.

Davis, R. B. (1975). Cognitive processes involved in solving simple algebraic equations. *Journal of Children's Mathematical Behavior*, **1** (3), 7–35.

Davis, R. B. (1984). *Learning mathematics: The cognitive science approach to mathematics education.* Norwood, NJ: Ablex Publishing Corporation.

Davis, R. B. (1990). The knowledge of cats: Epistemological foundations of mathematics education. Invited address to the Plenary Session of the Annual Meeting of the International Group for the Psychology of Mathematics Education, Oaxtepec, Mexico: CINVESTAV.

Davis, R. B.,,, C. A. Maher & N. Noddings (Eds.) (1990). *Constructivist views on the teaching and learning of mathematics.* (Journal for Research in Mathematics Education Monograph No. 4). Reston, VA: National Council of Teachers of Mathematics.

Deatsman, G. A. (1979). An experimental evaluation of retesting. *American Mathematical Monthly*, **86**, 51–53.

Deboer, G. E. (1984). A study of gender effects in the science and mathematics course–taking behavior of a group of students who graduated from college in the late 1970s. *Journal of Research in Science Teaching*, **21**, 95–103.

Dees, R. L. (1991). The role of cooperative learning in increasing problem–solving ability in a college remedial course. *Journal for Research in Mathematics Education*, **22**, 409–421.

Dessart, D. J. (1989). *A review and synthesis of research in mathematics education reported during 1987.* Columbus, OH: ERIC Clearinghouse for Science, Mathematics and Environmental Education.

Dew, K. M. H., J. P. Galassi & M. D. Galassi. (1984). Math anxiety: Relation with situational test anxiety, performance, physiological arousal, and math avoidance behavior. *Journal of Counseling Psychology*, **31** (4), 580–583.

Dick, T. (1988). Student use of graphical information to monitor symbolic calculations. (Working paper available from author: Dept. of Mathematics, Oregon State University, Corvallis, OR 97331).

Dick, T. P. & C. M. Patton. (1991). *Calculus* (Volume 1). Boston: PWS–Kent.

Dossey, J. A., I. V. S. Mullis, M. M. Lindquist & D. L. Chambers. (1988). *The mathematics report card: Are we measuring up? Trends and achievement based on the 1986 National Assessment of Educational Progress.* Princeton, NJ: Educational Testing Service.

Dougherty, B. J. (1990). Influences of teacher cognitive/conceptual levels on problem solving instruction. In G. Booker, P. Cobb & T. N. de Mendicuti (Eds.), *Proceedings of the Fourteenth Conference of the International Group for the Psychology of Mathematics Education*, **1**, 119–126. Mexico: CINVESTAV.

Douglas, R. G. (Ed.) (1986). *Toward a lean and lively calculus.* (MAA Notes No. 6). Washington DC: Mathematical Association of America.

Drew, D. (1983). Advanced ocularmetrics: Graphing multiple time series residual data. *Educational Evaluation and Policy Analysis*, **5** (1), 97–105.

Dreyfus, T. (1990). Advanced mathematical thinking. In P. Nesher, & J. Kilpatrick (Eds.), *Mathematics and cognition: A research synthesis by the International Group for the Psychology of Mathematics Education* (pp. 113–134). Cambridge: Cambridge University Press.

Dreyfus, T., & T. Eisenberg. (1982). Intuitive functional concepts: A baseline study on intuitions. *Journal for Research in Mathematics Education*, **13**, 360–380.

Dreyfus, T. & T. Eisenberg. (1983). The function concept in college students: Linearity, smoothness, and periodicity. Unpublished manuscript.

Dreyfus, T. & T. Eisenberg. (1990). On difficulties with diagrams: Theoretical issues. *Proceedings of the 14th International Conference for the Psychology of Mathematics Education*, **1**, (pp. 27–34). Oaxtepec, Mexico: CINVESTAV.

Driscoll, M. *Research within reach: Secondary school mathematics.* St. Louis and Washington DC: RDIS/NIE, 1983.

Dubinsky, E. (1986) Teaching mathematical induction I. *Journal of Mathematical Behavior*, **5**, 305–17.

Dubinsky, E. (1988). Teaching mathematical induction II. *Journal of Mathematical Behavior*, 284-304.

Dubinsky, E. (1988). On helping students construct the concept of quantification. In A. Borbas, *Proceedings of the Twelfth Annual Conference of the International Group for Psychology of Mathematics Education*, **1**, (pp. 255–262). Veszprem, HUNG: OOK Printing House.

Dubinsky, E. (Ed.) (1989). *UME Trends.* Lafayette, IN: American Mathematical Society.

Dubinsky, E. (October 1989). *Calculus, computers, and concepts.* Paper presented at the 19th Annual North Carolina Council of Teachers of Mathematics Conference, Raleigh, NC.

Dubinsky, E., F. Elterman & C. Gong. (June 1988). "The student's construction of quantification." *For the Learning of Mathematics*, **8** (2) 44-51

Dubinsky, E. & G. Harel. (1992). The nature of the process conception of function. In G. Harel and E. Dubinsky, (Eds.) *The concept of function: Aspects of epistemology and pedagogy.* MAA Notes and Reports, No. 25. Washington, DC: Mathematical Association of America , 85-106.

Dubinsky, E & U. Leron. (January, 1991). *Learning abstract algebra by programming in ISETL.* MAA Minicourse presented at the Annual Meeting of the Mathematical Association of America, San Francisco.

Dubinsky, E. & K. Schwingendorff. (1990). Calculus concepts and computers: Innovations in learning. In T. Tucker (Ed.), *Priming the calculus pump: Innovations and resources.* MAA Notes and Reports, No. 17 Washington, DC: Mathematical Association of America.

Dunn, R., R. I. Sklar, J. S. Beaudry & J. Bruno. (1990). Effects of matching and mismatching minority developmental college students' hemispheric preferences on mathematics scores. *Journal of Educational Research*, **83**, 283–288.

Eisenberg, T. (1981). Remedial mathematics and open admissions. *School Science and Mathematics*, **81**, 341–346.

Eisenberg, T. (1990). *On the development of a sense of functions.* Paper presented at the Conference on the Learning and Teaching of the Function Concept. Purdue University, West Lafayette, IN.

Eisenberg, T. & T. Dreyfus. (1985). Toward understanding mathematical thinking. In L. Streefland (Ed.), *Proceedings of the Ninth International Conference for the Psychology of Mathematics Education,* **1,** (pp. 241–246). Utrecht, The Netherlands: State University of Utrecht.

Eisenberg, T. & T. Dreyfus. (1986). On visual versus analytical thinking in mathematics, *Proceedings of the Tenth International Conference for the Psychology of Mathematics Education* (pp. 153–158). London: University of London Institute of Education.

Espigh, M. A. (1975). *A comparison of two guided discovery strategies and an expository strategy for teaching college freshmen proof of theorems based on the field axioms.* (Doctoral dissertation, The Florida State University, 1974). Dissertation Abstracts International, **35,** 934B.

Even, R. (1988). Pre–service teachers' conceptions of the relationships between functions and equations. In A. Borbas (Ed.), *Proceedings of the Twelfth Annual Conference of the International Group for the Psychology of Mathematics Education,* **1,** (pp. 304–311). Veszprem, HUNG: OOK Printing House.

Even, R., G. Lappan & W. Fitzgerald. (1988). Pre–service teachers' conceptions of the relationship between functions and equations. In M. Behr, C. Lacampagne & M. M. Wheeler (Eds.), *Proceedings of the Tenth Annual Meeting of the North American Chapter of the International Group for the Psychology of Mathematics Education.* DeKalb, IL: Northern Illinois University.

Even, R. (1990). The two faces of the inverse function: Prospective teachers' use of "undoing." In G. Booker, P. Cobb & T. N. de Mendicuti (Eds.), *Proceedings of the Fourteenth Conference of the International Group for the Psychology of Mathematics Education,* **1,** (pp. 37–44), Mexico: CINVESTAV.

Eylon, B. S. & M. C. Linn. (1988). Learning and instruction: An examination of four research perspectives in science education. *Review of Educational Research,* **58,** 251–301.

Fawcett, H. P. (1938). The nature of proof: *The thirteenth yearbook of National Council of Teachers of Mathematics.* Reston, VA: NCTM.

Fennema, E. (1989). The study of affect and mathematics: A proposed generic model for research, in *Affect and mathematical problem solving.* NY: Springer–Verlag.

Fennema, E. & M. Meyer. (1989). Gender, equity and mathematics. In W.G. Secada (Ed.), *Equity in education.* NY: The Falmer Press, 146–157.

Fennema, E. & G. Leder. (1990). *Mathematics and gender.* NY: Teachers College Press.

Ferrini–Mundy, J. & K. Graham. (1991). An overview of the calculus curriculum reform effort: Issues for learning, teaching, and curriculum development. *American Mathematical Monthly,* **98,** (7), 627-635.

Ferrini–Mundy, J., & Lauten.D. (1993). Teaching and learning calculus. In P. S. Wilson (Ed.), *Research ideas for the classroom,* (155-176). NY: Macmillan.

Fetta, I. & J. G. Harvey. (1990a). Technology is changing tests and testing. *UME Trends,* **1,** (6), 1; 4.

Fetta, I. & J. G. Harvey. (1990b). Computerized mathematics testing. *UME Trends,* **2,** (1), 1; 6;7.

Feurzeig, W., S. Wight & J. Richards. (1988). Pluribus: A visual programming environment for education and research, *Proceedings of the IEEE Workshop on Language Automation,* College Park, MD.

Fischer, R. (June, 1988). Didactic, mathematics, and communication. *For the Learning of Mathematics.*

Flener, F. O. (1990). Can teachers evaluate problem solving ability? *Proceedings of the 14th International Conference for the Psychology of Mathematics Education,* **1,** (pp. 127–134). Oaxtepec, Mexico: CINVESTAV.

Fox, L. H., E. Fennema & J. Sherman. (1977). *Women and mathematics: Research perspectives for change.* National Institute of Education.

Fraleigh, J., B. & L. I. Pakula. (1986). *Exploring calculus with the IBM PC*. NY: Addison–Wesley Publishing Company.

Franklin, J. & A. Daoud. (1988). *Introduction to proofs in mathematics*. Sydney: Prentice Hall.

Frazier Sr., R. C. (1970). *A comparison of an implicit and two explicit methods of teaching mathematical proof via abstract groups using selected rules of logic*. (Doctoral dissertation, The Florida State University, 1969). Dissertation Abstracts International, **30**, 5317A.

Friedman, L. (1989). Mathematics and the gender gap: A meta–analysis of recent studies on sex differences in mathematical tasks. *Review of Educational Research*, **59**, 185–213.

Fullilove, R. E. & P. U. Treisman. (1990). Mathematics achievement among African–American undergraduates at the University of California at Berkeley: An evaluation of the mathematics workshop program. *Journal of Negro Education*, **59**, 463–478.

Gardner, M. (1979). *Mathematical circus*. New York: Alfred Knopf.

Gardner, M. (1967). *The Scientific American book of mathematical puzzles and diversionss*. NY:: Simon and Schuster.

Gentry, W. M. & R. Underhill. (1987). A comparison of two palliative methods of intervention for the treatment of mathematics anxiety among female college students. In J. C.Bergeron, N. Herscovics & C. Kieran (Eds.), *Proceedings of the Eleventh International Conference for the Psychology of Mathematics Education*, **1**, (pp. 99–105). Montreal, Canada.

Gerber, H. C. (1972). *An investigation of the effects of programmed instruction in logical inferences upon college students' ability to learn proof writing*. (Doctoral dissertation, The Florida State University, 1971). Dissertation Abstracts International, **34**, 4908A.

Ginsburg, H. (Ed.). (1983). *The development of mathematical thinking*. NY: Academic Press.

Gliner, G. S. (1989). College students' organization of math word problems in relation to success in problem solving. *School Science and Mathematics*, **89**, 392–404.

Gliner, G. S. (1991). College students' organization of math word problem solving in terms of mathematical structure versus surface structure. *School Science and Mathematics*, **91**, 105–110.

Goldberg, D. J. (1975). *The effects of training in heuristic methods on the ability to write proofs in number theory*. (Doctoral dissertation, Columbia University, 1974). Dissertation Abstracts International, **35**, 4989B.

Goldenberg, E. P. (1987). Believing is seeing: How preconceptions influence the perception of graphs. In J. C. Bergeron, N. Herscovics & C. Kieran (Eds.), *Proceedings of the Eleventh International Conference for the Psychology of Mathematics Education*, **1**, (pp. 197–203). Montreal, Canada.

Goldenberg, E. P., P. Lewis & J. O'Keefe. (1992) Dynamic representation in the development of a process understanding of functions. In G. Harel and E. Dubinsky, (Eds.) *The concept of function: Aspects of epistemology and pedagogy*. MAA Notes and Reports, No. 25. Washington, DC: Mathematical Association of America, 235-260.

Graham, K. G. & J. Ferrini–Mundy. (March, 1989). *An exploration of student understanding of central concepts in calculus*. Paper presented at the Annual Meeting of the American Educational Research Association, San Francisco.

Green, L. T. (1990). Test anxiety, mathematics anxiety, and teacher comments: Relationships to achievement in remedial mathematics classes. *Journal of Negro Education*, **59**, 320–335.

Greenes, C. & W. Fitzgerald (1991). Mathematical performance of non-math majors at the college level. In R. G. Underhill (Ed.), *Proceedings of the Thirteenth Annual Meeting, North American Chapter of the International Group for the Psychology of Mathematics Education*, **1**, (pp. 98-104). Blacksburg, VA: VPI

Grossman, A. S. (1983). Decimal notation: An important research finding. *Arithmetic Teacher*, **30**, 32–33.

Hall, H. S. & S. R. Knight. (1887). *Higher algebra*. MacMillan and Co., London.

Hall, R. M. & Sandler, B. R. (1982). *The classroom climate: A chilly one for women?* Washington, DC: Project on the Status and Education of Women.

Hamblin, R. L., C. Hathaway, & J. S. Wodarski. (1971). Group contingencies, peer tutoring, and accelerating academic achievement. In E. Ramp and W. Hopkins (Eds.), *A new direction for education: Behavior analysis*. Lawrence, KS: The University of Kansas, Department of Human Development, 41–53.

Hanna, Gila. (1983). *Rigorous proof in mathematics education.* Ontario: Ontario Institute for Studies in Education Press.

Hanna, Gila. (1989). More than formal proof. *Learning Mathematics— An International Journal of Mathematics Education,* **9,** 20–23.

Hanna, Gila & I. Winchester (Eds.) (1990). *Creativity, thought and mathematical proof.,* special issue of *Interchange,* 21.

Hansen, R. S., J. McCann, & J. L. Meyers. (1985). Rote versus conceptual emphases in teaching elementary probability. *Journal for Research in Mathematics Education,* **16,** 364–374.

Hardiman, P. T. (1984). Usefulness of a balance model in understanding the mean. *Journal of Educational Psychology,* **76,** 792–801.

Harel, G. (1989a). Learning and teaching linear algebra: Difficulties and an alternative approach to visualizing concepts and processes. *Focus on Learning Problems in Mathematics,* **11,** 139–148.

Harel, G. (1989b). Applying the principle of multiple embodiments in teaching linear algebra: Aspects of familiarity and mode of representation. *School Science and Mathematics,* **89,** 49–57.

Harel, G. & E. Dubinsky. (1991). The development of the concept of function by preservice secondary teachers from action conception to process conception. In F. Furinghetti (Ed.), *Proceedings of the Fifteenth Conference of the International Group for the Psychology of Mathematics Education,* **2,** (pp. 133–140). Genova, Italy.

Harel, G. & E. Dubinsky. (1992). *The concept of function: Aspects of epistemology and pedagogy.* Washington, DC: Mathematical Association of America.

Harel, G. (January, 1991). *On the construction of knowledge in mathematics: Formation of entities, abstraction, and generalization.* Paper presented at the Annual Meeting of the American Mathematical Society, San Francisco.

Harel, G. & J. J. Kaput. (1990). The role of conceptual entities in learning mathematical concepts at the undergraduate level. In G. Booker, P. Cobb & T. N. de Mendicuti (Eds.), *Proceedings of the Fourteenth Conference of the International Group for the Psychology of Mathematics Education,* **1,** (pp. 53–60). Mexico: CINESTAV.

Hart, L. C. (1991). Assessing teacher change in the Atlanta Math Project. In R. G. Underhill (Ed.), *Proceedings of the Thirteenth Annual Meeting, North American Chapter of the International Group for the Psychology of Mathematics Education,* **2,** (pp. 78–84). Blacksburg, VA: VPI.

Harvey, Wayne. (1982). *Success and failure in problem solving: An investigation of mental processing.* (Doctoral dissertation, University of California, Berkeley, 1981). Dissertation Abstracts International, **42,** 4916B.

Hativa, N. (1983). What makes mathematics lessons easy to follow, understand, and remember? *Two–Year College Mathematics Journal,* **14,** 398–406.

Hativa, N. (1985). A study of the organization and clarity of mathematics lessons. *International Journal of Mathematics Education in Science and Technology,* **16,** 89–99.

Heid, M. K. (1988). Resequencing skills and concepts in applied calculus using the computer as a tool. *Journal for Research in Mathematics Education,* **19,** 3–25.

Heid, M. K. (1990). Uses of technology in prealgebra and beginning algebra. *Mathematics Teacher,* **83,** 194–198.

Herscovics, N. (1989). Cognitive obstacles encountered in the learning of algebra. In S. Wagner & C. Kieran (Eds.), *Research issues in the learning and teaching of algebra* (pp. 60–86). National Council of Teachers of Mathematics and Lawrence Erlbaum Associates.

Higginson, W. (1982). Symbols, icons, and mathematical understanding. *Visible Language,* **16,** 239–248.

Hirsch, C. R., S. F. Kapoor & R. A. Laing. (1983). Homework assignments, mathematical ability, and achievement in calculus. *Mathematics and Computer Education,* **17,** 51–57.

Hitt, F. (1989). Construction of functions, contradiction and proof. In G. Vergnaud, J. Rogalski & M. Artigue (Eds.), *Proceedings of the Thirteenth International Conference of Psychology of Mathematics Education,* **2,** (pp. 107–114). Paris, FR: G.R. Didactique.

Hoffer, A. (1983) Van Hiele–based research. In R. Lesh & M. Landau (Eds.), *Acquisition of mathematics concepts and processes.* NY: Academic Press,.

Honsberger, R. (1973) *Mathematical gems, dolciani mathematical expositions.* Washington, DC: The Mathematical Association of America,.

Honsberger, R. (1970) *Ingenuity in mathematics: New mathematical library.* Washington, DC: The Mathematical Association of America,.

Howell, Edgar N. & R. Melander (1967). College students' ability to prove mathematical theorems with and without training in inference patterns. *The Journal of Experimental Education*, 35, 58–65.

Hsu, T. & M. D.Shermis. (1989). The development and evaluation of a microcomputerized adaptive placement testing system for college mathematics. *Journal of Educational Computing Research*, 5, 473–485.

Jackson, A. (1989a). NSF calculus projects: Progress report. *UME Trends*, 1 (2), 1; 5.

Jackson, A. (1989b). NSF calculus projects: Progress report. *UME Trends*, 1 (3), 1; 5.

Janvier, C. (Ed.) (1987). *Problems of representation in the teaching and learning of mathematics.* Hillsdale, NJ: Lawrence Erlbaum Associates.

Johnson, D. W., R. T. Johnson & L. Scott. (1978). The effects of cooperative and individualized instruction on student attitudes and achievement. *Journal of Social Psychology*, 104, 207–216.

Johnston, W. B. & A. E. Packer (Eds.) (1987). *Workforce 2000: Work and workers for the 21st century.* Indianapolis, IN: Hudson Institute.

Jurdak, M. (1991). Teachers' conceptions of math education and the foundations of mathematics. In F. Furinghetti (Ed.), *Proceedings of the Fifteenth Conference of the International Group for the Psychology of Mathematics Education*, 2, (pp. 221–228). Genova, Italy.

Kaput, J. J. (1979). Mathematics and learning: Roots of epistemological status. In J. Clement & J. Lochhead (Eds.), *Cognitive process instruction.* (pp. 289–303). Philadelphia, PA: Franklin Institute Press.

Kaput, J. J. (1987) Representation systems and mathematics. In C. Janvier (Ed.), *Problems of representation in the teaching and learning of mathematics.* Hillsdale, NJ: Lawrence Erlbaum Associates.

Kaput, J. J. (1989). Linking representations in the symbol systems of algebra. In C. Kieran & S. Wagner (Eds.), *Research issues in the learning and teaching of algebra.* (pp. 167-194) Reston, VA: National Council of Teachers of Mathematics and Lawrence Erlbaum Associates,.

Kaput, J. J. (January, 1991). *A framework for understanding the learning and use of mathematical notations.* Paper presented at the Annual Meeting of the American Mathematical Society, San Francisco.

Keith, S. Z. & Phillip (Eds.) (1989). *Proceedings of the national conference on women in mathematics and science*, St. Cloud State University, St. Cloud, MN.

Kenschaft, P. C. (Ed.), (1990) *Winning women into mathematics.* Washington, DC: The Mathematical Association of America,.

Keny, S. V. (1990). Bridging the gap between lower and upper division math. *UME Trends*, 2, 3, p.2.

Kieran, C. (1989). The early learning of algebra: A structural perspective. In S. Wagner & C. Kieran (Eds.), *Research issues in the learning and teaching of algebra* (pp. 167–194). National Council of Teachers of Mathematics and Lawrence Erlbaum Associates.

Kiser, L. (1990). Interaction of spatial visualization with computer–enhanced and traditional presentations of linear absolute–value inequalities. *Journal of Computers in Mathematics and Science Teaching*, 10, 85–96.

Koblitz, A. H. (1983). *A convergence of lives, Sofia Kovalevskaia: Scientist, writer, revolutionary.* Boston: Birkhauser.

Krutetskii, V. A. (1976). *The psychology of mathematical abilities in schoolchildren.* J. Kilpatrick & I. Wirszup (Eds.). (J. Teller, Tr.), University of Chicago Press: Chicago.

Kuhn, T. S. (1970). *The structure of scientific revolutions* (2nd ed.). University of Chicago Press: Chicago.

Lampert, M. (1988). The teacher's role in reinventing the meaning of mathematical knowing in the classroom. In M. Behr, C. Lacampagne, & M.M. Wheeler (Eds.), *Proceedings of the Tenth Annual Meeting of the North American Chapter of the International Group for the Psychology of Mathematics Education.* DeKalb, IL: Northern Illinois University.

Land, M. L. & L. R. Smith. (1979). Effects of a teacher clarity variable on student achievement. *Journal of Educational Research*, 72, 196–197.

Lawson, M. J. (1990). The case for instruction in the use of general problem–solving strategies in mathematics teaching: A comment on Owen and Sweller. *Journal for Research in Mathematics Education*, **21**, 403–410.

Lazarus, R. S. (1986). Cognition and emotion from the ret viewpoint. (M.E. Bernard and R. Digiuseppe, Eds.) *Inside rational emotive therapy*, NY: Academic Press.

Lazarus, R. S., A..D. Kanner & S. Folkman (1980). Emotions: A cognitive–phenomenological analysis. In R. Plutchik and H. Kellerman (Eds.), *Emotion, theory, research and experience*, **1**, NY: Academic Press.

Lean, G. & M. A. Clements. (1981). Spatial ability, visual imagery, and mathematical performance. *Educational Studies in Mathematics*, **12**, 267–299.

Leinhardt, G., O. Zaslavsky & M. K.Stein. (1990). Functions, graphs, and graphing: Tasks, learning, and teaching. *Review of Educational Research*, **60** (1), 1–64.

Leitzel, J. R. (1983). Improving school–university articulation in Ohio. *Mathematics Teacher*, **76**, 610–616.

Leitzel, J. R. (Ed.).(1990) *A call for change: Recommendations for the mathematical preparation of Teachers of Mathematics*, Committee On the Mathematical Education of Teachers, Washington, DC: The Mathematical Association of America,,

Leron, U. (1983). Structuring mathematical proofs. *The American Mathematical Monthly*, **90**, 174–185.

Leron, U. *Learning abstract algebra via programming in ISETL*. (January, 1991) Paper presented at the Annual Meeting of the American Mathematical Society, San Francisco,.

Leron, U. and E. Dubinsky.(1993) *Learning abstract algebra.with ISETL*, Springer-Verlag: NY.

Lesh, R. & M. Landau (Eds.) (1983). *Acquisition of mathematics concepts and processes*. NY: Academic Press.

Lesh, R. (1985). A. Conceptual analyses of problem–solving performance. In E. A. Silver (Ed.), *Teaching and learning mathematical problem solving: Multiple research perspectives*. Hillsdale, NJ: Lawrence Erlbaum Associates.

Lester Jr., F. K. (1985). Methodological considerations in research on mathematical problem–solving instruction. In E. A. Silver (Ed.), *Teaching and learning mathematical problem solving: Multiple research perspectives*. Hillsdale, NJ: Lawrence Erlbaum Associates, Publishers.

Levin, G. (1990) *ISETL: Version 3.0*, Aug. Available from R. Pullins, West Educational Publishing, Hamilton Gateway Building, 5 Market Square, Amesbury, MA 01913.

Lewis, M. W. & J. R. Anderson. (1985). Discrimination of operator schemata in problem solving: Learning from examples. *Cognitive Psychology*, **17**, 26–65.

Linn, M. C. & J. S. Hyde. (1989). Gender, mathematics, and science. *Educational Researcher*, **18**, 17–19; 22–27.

Livingston, C. & H. Borko. (1990). High school mathematics review lessons: Expert–novice distinctions. *Journal for Research in Mathematics Education*, **21**, 372–387.

Llabre, M. M., & E. Suarez. (1985). Predicting math anxiety and course performance in college men and women. *Journal of Counseling Psychology*, **32** (2), 283–287.

Lohnas, E. K. (1990). LARC project for the development of classroom–based research and learning outcomes data base within a consortium context. *CACC News*, **35** (1), 9.

Lovell, K. (1971). The development of the concept of mathematical proof in abler pupils. In M. F. Rosskopf, L. P. Steffe, & S. Tabac (Eds.), *Piagetian cognitive–development research and mathematical education*. Reston, VA: NCTM.

Lucas, J. F. (1980). An exploratory study on the diagnostic teaching of heuristic problem–solving strategies in calculus. In J. G. Harvey and T. A. Romberg (Eds.), *Problem–solving studies in mathematics*. Madison, WI: Wisconsin Research and Development Center for Individualized Schooling Monograph Series.

Macey, W. T. (1971). *An investigation of the effect of prior instruction of selected topics of logic on the understanding of the limit of a sequence*. (Doctoral dissertation, The Florida State University, 1970). *Dissertation Abstracts International*, **31**, 5490B.

MacGregor, S. K., J. Z. Shapiro & R. Niemier. (1989). Effects of a computer–augmented learning environment on mathematics achievement for students with differing cognitive style. *Journal of Educational Computing Research*, **4**, 453–465.

Malone, J. A., et. al. (1980) Measuring problem–solving ability. In Stephen Krulik (Ed.), *Problem solving in school mathematics: 1980 Yearbook of the National Council of Teachers of Mathematics.* Reston, VA: NCTM.

Mandler, G. (1989). Affect and learning: Causes and consequences of emotional interactions and affect and learning: Reflections and prospects, In D. McLeod & V. Adams, (Eds.) *Affect and mathematical problem solving: A new perspective.* NY: Springer–Verlag.

Mandler, G. (1975). *Mind and emotion.* NY: Wiley.

Mangan, K. S. (May, 1987). Undergraduates, professors collaborate on research at more and more colleges. *Chronicle of Higher Education*, 1, 26.

Markovits, Z., B. Eylon & M. Bruckheimer. (1986). Functions today and yesterday. *For the Learning of Mathematics*, **6** (2), 18–28.

Martin, G. W. & G. Harel. (1989). Proof frames of preservice elementary teachers. *Journal for Research in Mathematics Education*, **20**, 41–51.

Martin, G. W. & Harel, G. (1989). The role of the figure in students' concepts of geometric proof. In G. Vergnaud, J. Rogalski & M. Artigue (Eds.), *Proceedings of the Thirteenth International Conference of Psychology of Mathematics Education*, **2**, (pp. 266–273). Paris, FR: G.R. Didactique.

Marty, R. H. (1990). An alternative instructional approach to transition courses for mathematics majors. *UME Trends*, **1** (3), p. 2.

Mason, J., L. Burton & K. Stacey (1982). *Thinking mathematically*, Addison–Wesley: London.

Mathews, J. H. (In press). Computer algebra systems approach to teaching Taylor polynomials. *College Mathematics Journal.*

Matz, M. (1979). *Towards a process model for high school algebra errors.* (Working paper No. 181). Cambridge: Massachusetts Institute of Technology, Artificial Intelligence Laboratory.

Matz, M. (1983). Towards a computational theory of algebra competence. *Journal of Mathematical Behavior*, **3**, 93–166.

Mayer, R. E. (1977). Different rule systems for counting behavior acquired in meaningful and rote contexts of learning. *Journal of Educational Psychology*, **69**, 537–546.

McCoy, R. E. (1972). *A study of the effects of three different strategies of proof instruction and background factors of elementary education majors for success in constructing deductive proof in mathematics.* (Doctoral dissertation, The Pennsylvania State University, 1971). Dissertation Abstracts International, **32**, 5091A.

McKnight, C. C., Crosswhite, F. J., Dossey, J. A., Kifer, E., Swafford, J. O., Travers, K. J., & Cooney, T. J. (1987). *The underachieving curriculum: Assessing U.S. school mathematics from an international perspective.* Champagne, IL: Stipes Publishing Co.

McLeod, D. B. (1985). Affective influences on mathematical problem solving. In L. Streefland (Ed.), *Proceedings of the Ninth International Conference for the Psychology of Mathematics Education*, **1**, (pp. 259–269). Utrecht, The Netherlands: State University of Utrecht.

McLeod, D. B. (1987). Beliefs, attitudes, and emotions: Affective factors in mathematics learning. In J. C.Bergeron, N. Herscovics, & C.Kieran (Eds.), *Proceedings of the Eleventh International Conference for the Psychology of Mathematics Education*, **1**, (pp. 170–182). Montreal, Canada.

McLeod, D. B. & V. M. Adams (Eds.) (1989). *Affect and mathematical problem solving: A new perspective*, NY: Springer–Verlag.

Mestre, J. P., W. J. Gerace & J. Lochhead. (1982). The interdependence of language and translational math skills among bilingual Hispanic engineering students. *Journal of Research in Science Teaching*, **19**, 399–410.

Meyer, L. J. (1983). *Teaching problem solving in a college level general education mathematics class.* (Doctoral dissertation, The University of Iowa, 1982). Dissertation Abstracts International, **43**, 2584A.

Monk, G. S. (1987). *Students' understanding of functions in calculus courses,* Unpublished manuscript.

Monk, G. S. (1992). *Students' understanding of a function given by a physical model.* In G. Harel and E. Dubinsky, (Eds.) *The concept of function: Aspects of epistemology and pedagogy.* MAA Notes and Reports, No. 25. Washington, DC: Mathematical Association of America , 175-194.

Moore, R. C. (1991). A concept–understanding scheme and the learning of proof. In R. G. Underhill (Ed.), *Proceedings of the Thirteenth Annual Meeting, North American Chapter of the International Group for the Psychology of Mathematics Education*, **2**, (pp. 210–216). Blacksburg, VA: VPI & SU.

Morgan, W. H. (1972). *A study of the abilities of college mathematics students in proof–related logic.* (Doctoral dissertation, University of Georgia, 1971). Dissertation Abstracts International, **32**, 4081B.

Movshovitz–Hadar, N. (1988). Stimulating presentation of theorems followed by responsive proofs. *For the Learning of Mathematics*, 8, 12–19.

Muench, D. (March, 1990). Review of *ISETL: Interactive Set Language. Notices of the American Mathematical Society.*

Mura, R. (1987). Sex related differences in expectations of success in undergraduate mathematics. *Journal for Research in Mathematics Education*, **18**, 15–24.

National Council of Teachers of Mathematics (1989). *Curriculum and evaluation standards for school mathematics.* Reston, VA: NCTM.

National Council of Teachers of Mathematics (1991). *Professional standards for teaching mathematics.* Reston, VA: NCTM.

National Research Council. (1986). *Mathematical sciences: A unifying and dynamic resource.* Washington, D.C.: National Academy Press.

National Research Council. (1989). *Everybody counts: A report to the nation on the future of mathematics education.* Washington, DC: National Academy Press.

National Research Council. (1991). *Moving beyond myths: Revitalizing undergraduate mathematics.* Washington, DC: National Academy Press.

National Science Board Commission on Precollege Education in Mathematics, Science and Technology. (1983). *Educating Americans for the 21st century: A plan of action for improving mathematics, science and technology education for all American elementary and secondary students so that their achievement is the best in the world by 1995.* Washington, DC: National Science Board.

National Science Foundation. (1988). *Women and minorities in science and engineering.* Washington, DC: Author.

Noddings, N. (1990). Constructivism in mathematics education. In R. B. Davis, C. A. Maher, & N. Noddings, (Eds.) *Constructivist views on the teaching and learning of mathematics.* (Journal for Research in Mathematics Education Monograph No. 4). Reston, VA: National Council of Teachers of Mathematics.

Norman, F. A. (1986). Students' unitizing of variable complexes in algebraic and graphical situations. In G. Lappan and R. Even (Eds.), *Proceedings of the Eighth Annual Meeting of PME–NA*, (pp. 102–107). Michigan State University, East Lansing.

Norman, F. A. (1987). *An examination of students' uses of unitizing strategies in algebraic and graphical contexts.* Unpublished doctoral dissertation, University of Georgia: Athens, GA.

Nummedal, S. G. & F. P. Collea, (1981). Field independence, task ambiguity, and performance on a proportional reasoning task. *Journal of Research in Science Teaching*, **18**, 255–260.

Ohlson, E. L. & L.Mein. (1977). The difference in level of anxiety in undergraduate mathematics and nonmathematics majors. *Journal for Research in Mathematics Education*, 8, 48–56.

Orton, A. (1983a). Students' understanding of differentiation. *Educational Studies in Mathematics*, **15**, 235–250.

Orton, A. (1983b). Students' understanding of integration. *Educational Studies in Mathematics*, **14**, 1–18.

Owen, E. & J. Sweller. (1989) Should problem solving be used as a learning device in mathematics? *Journal for Research in Mathematics Education*, **20**, 322–328.

Palmiter, J. (1990). Using computer algebra systems in calculus. In F. Demana & J. Harvey (Eds.), *Proceedings of the Conference on Technology in Collegiate Mathematics*, (pp. 68–70). Menlo Park, CA: Addison–Wesley Publishing Company.

Palmiter, J. (1991). Effects of computer algebra systems on concept and skill acquisition in calculus. *Journal for Research in Mathematics Education*, **22**, 151–156.

Pascarella, E. T. (1977). Student motivation as a differential predictor of course outcomes in personalized system of instruction and conventional instructional methods. *Journal of Educational Research*, **71**, 21–26.

Pascarella, E. T. (1985). College environmental influences on learning and cognitive development: A critical review and synthesis. In J. C. Smart (Ed.), *Higher education: Handbook of theory and research,* **1,** NY: Agathon.

Patterson, W. H. (1970). *The development and testing of a discovery strategy in mathematics involving the field axioms.* (Doctoral dissertation, The Florida State University, 1969). Dissertation Abstracts International, **30,** 5599B.

Perry, W. G. (1970). *Forms of intellectual and ethical development in college: A scheme.* NY: Holt, Rinehart and Winston.

Peterson, P. L., & T. C. Janicki. (1979). Individual characteristics and children's learning in large–group and small–group approaches. *Journal of Educational Psychology,* **71,** 677–687.

Pimm, D. (1987). *Speaking mathematically: communication in mathematics classrooms.* London: Routledge.

Poincaré, H. (1914). *Science and method.* NY: Dover.

Polya, G. (1954). *How to solve it.* Princeton, NJ: Princeton University Press.

Polya, G. (1957) *Mathematics and plausible reasoning (Volume I: Induction and analogy in mathematics; Volume II: Patterns of plausible inference).* Princeton, New Jersey: Princeton University Press.

Rachlin, S.L. (1987). *Algebra from x to y: A process approach for developing the concepts and generalizations of algebra.* Curriculum materials, University of Hawaii.

Ralston, A. (1990). The effect of technology on teaching college mathematics. In F. Demana & J. Harvey (Eds.), *Proceedings of the Conference on Technology in Collegiate Mathematics,* (pp. 78–82). Menlo Park, CA: Addison–Wesley Publishing Company.

Reed, S. K. (1989). Constraints on the abstraction of solutions. *Journal of Educational Psychology,* **81,** 532–540.

Reisel, R. B. (1982). How to construct and analyze proofs — a seminar course. *The American Mathematical Monthly,* **89,** 490–492.

Resek, D. & W. H. Rupley, (1980). Combatting 'mathophobia' with a conceptual approach toward mathematics. *Educational Studies in Mathematics,* **11,** 423–441.

Resnick, H., J. Viehe & S. Segal. (1982). Is math anxiety a local phenomenon? A study of prevalence and dimensionality. *Journal of Counseling Psychology,* **29,** 39–47.

Resnick, L. B., & S. F. Omanson. (1987). Learning to understand arithmetic. In R. Glaser (Ed.) *Advances in instructional psychology,* **3,** Hillsdale, NJ: Erlbaum.

Rogers, P. (1988). Student–sensitive teaching at the tertiary level: A case study. In A. Borbas (Ed.), *Proceedings of the Twelfth Annual Conference of the International Group for the Psychology of Mathematics Education,* **2,** (pp. 536–543). Veszprem, HUNG: OOK Printing House.

Rosamond, F.& J. Poland (June, 1985). The role of feelings in learning mathematics. *CMESG/GCEDM, Proceedings,* Université Laval.

Ross, Peter. (1980). *Student difficulties in solving calculus word problems.* (Doctoral dissertation, University of California, Berkely. Dissertation Abstracts International, 1981, **41,** 3465A.

Russell, S.J. & R. B. Corwin. (1991). Talking Mathematics: "Going Slow" and "Letting Go." In R. G. Underhill (Ed.), *Proceedings of the Thirteenth Annual Meeting, North American Chapter of the International Group for the Psychology of Mathematics Education,* **2,** (pp. 175–181). Blacksburg, VA: VPI & SU.

St. Andre, R. J. & D. D. Smith (1978). "Proofs" to grade. *The American Mathematical Monthly,* **85,** 493–494.

Sasser, J. E. (1990/91). The effect of using computer tutorials as homework assignments on the mathematics achievement of elementary education majors. *Journal of Computers in Math and Science Teaching,* **10,** 95–102.

Schoen, H. L. (1976). Self–paced mathematics instruction: How effective has it been in secondary and postsecondary school? *Mathematics Teacher,* **69,** 352–357.

Schoenfeld, A. H. (1979). Explicit heuristic training as a variable in problem–solving performance. *Journal for Research in Mathematics Education,* **10,** 173–187.

Schoenfeld, A. H. (1982). Measures of problem–solving performance and of problem–solving instruction. *Journal for Research in Mathematics Education,* **13,** 31–49.

Schoenfeld, A. H. (1983a). *Problem solving in the mathematics curriculum: A report, recommendations, and an annotated bibliography.* Washington, DC: Mathematical Association of America.

Schoenfeld, A. H. (1983b). Beyond the purely cognitive: Belief systems, social cognitions, and metacognitions as driving forces in intellectual performance. *Cognitive Science,* **7,** 329–363.

Schoenfeld, A. H. (1983c). Episodes and executive decisions in mathematical problem solving. In R. Lesh & M. Landau (Eds.), *Acquisition of mathematics concepts and processes.* NY: Academic Press.

Schoenfeld, A. H. (1985). *Mathematical problem solving.* Academic Press: Orlando.

Schoenfeld, A. H. (1986). On having and using geometrical knowledge. In J. Hiebert (Ed.), *Conceptual and procedural knowledge: The case of mathematics.* Hillsdale, NJ: Lawrence Erlbaum Associates.

Schoenfeld, A. H. (Ed.) (1987). *Cognitive science and mathematics education.* Hillsdale, NJ: Lawrence Erlbaum Associates.

Schoenfeld, A., J. P. Smith & A. Arcavi. (1993). Learning: The microgenetic analysis of one student's evolving understanding of a complex subject matter domain. In R. Glaser (Ed.), *Advances in instructional psychology.* **4,** Hillsdale, NJ: Lawrence Erlbaum Associates, 55-175.

Schwartz, J. and M. Yerushalmy (1992). Getting students to function in and with algebra. In G. Harel and E. Dubinsky, (Eds.) *The concept of function: Aspects of epistemology and pedagogy.* MAA Notes and Reports, No. 25. Washington, DC: Mathematical Association of America, 261-289.

Schwartz, J., R. Dewar, E. Dubinsky & E. Schonberg (1986). *Programming with Sets: An Introduction to SETL.* NY: Springer–Verlag.

Selden, A. & J. Selden (1987). Errors and misconceptions in college level theorem proving, *Proceedings of the Second International Seminar on Misconceptions and Educational Strategies in Science and Mathematics,* **3.** Cornell University.

Selden, J., A. Mason & A. Selden. (1989). Can average calculus students solve nonroutine problems? *Journal of Mathematical Behavior,* **8,** 45–50.

Senk, S. L. (1985). How well do students write geometry proofs? *Mathematics Teacher,* **78,** 448–456.

Sfard, A. (1992). Operational origins of mathematical objects and the quandary of reification — the case of function. In G. Harel and E. Dubinsky, (Eds.) *The concept of function: Aspects of epistemology and pedagogy.* MAA Notes and Reports, No. 25. Washington, DC: Mathematical Association of America.

Shaughnessy, J. M. (1977). Misconceptions of probability: An experiment with a small–group, activity–based, model building approach to introductory probability at the college level. *Educational Studies in Mathematics,* **8,** 295–316.

Shavelson, R.J. (1988). Contributions of educational research to policy and practice: Constructing, challenging, changing cognition. *Educational Researcher,* **17,** (7), 4–11.

Siegel, R. G., J. P. Galassi & W. B. Ware (1985). A comparison of two models for predicting mathematics performance: Social learning versus math aptitude–anxiety. *Journal of Counseling Psychology,* **32,** (4), 531–538.

Silver, E. A. (Ed.) (1985a). *Teaching and learning mathematical problem solving: Multiple research perspectives.* Hillsdale, NJ: Lawrence Erlbaum Associates, Publishers.

Silver, E. (1985b). A. Research on teaching mathematical problem solving: some underrepresented themes and needed directions. In Edward A. Silver (Ed.), *Teaching and learning mathematical problem solving: multiple research perspectives.* Hillsdale, New Jersey: Lawrence Erlbaum Associates, Publishers.

Silver, E. A. (1988). "NCTM Curriculum and Evaluation Standards for School Mathematics: Responses from the Research Community." *Journal for Research in Mathematics Education,* **19,** No. 4, 338–344.

Silver, E. A. (1989). Teaching and assessing mathematical problem solving: Toward a research agenda. In R.I.Charles & E. A. Silver,(Eds.) (1989). *Research agenda for mathematics education: The teaching and assessing of mathematical problem solving,* (pp. 273–282). Reston, VA: National Council of Teachers of Mathematics.

Silver, E. (1990). Contribution of research to practice: Applying findings, methods, and perspectives. In T. Cooney and C. Hirsch (Eds.), *Teaching and learning mathematics in the 1990s: 1990 Yearbook,* Reston, VA: NCTM.

Silver, Edward A. & W. Metzger (1989). Expertise is a matter of taste as well as competence: Aesthetic monitoring in mathematical problem solving, In D. McLeod and V. Adams (Eds.) *Affect and mathematical problem solving*, NY: Springer–Verlag.

Skemp, R. (Ed.). (1982). Understanding the symbolism of mathematics. *Visible Language*, **16**, *passim*.

Skemp, R. R. (1986). *The psychology of learning mathematics*, second edition. Middlesex, England: Penguin Books Ltd.

Slavin, R. E., & N. Karweit (1981). Cognitive and affective outcomes of an intensive student team learning experience. *Journal of Experimental Education*, **50**, 29–35.

Slavin, R. E. & N. Karweit (1982). *Student teams and mastery: A factorial experiment in urban Math 9 classes*. Paper presented at the annual convention of the American Educational Research Association, NY.

Smith, J., A. Arcavi, & A. H.Schoenfeld (1989). Learning y–inter*cept*: Assembling the pieces of an "atomic" concept. In G. Vergnaud, J. Rogalski, & M. Artigue (Eds.), *Proceedings of the Thirteenth International Conference for the Psychology of Mathematics Education*, **3**, (pp. 174–181). Paris, FR: G.R. Didactique.

Smith, L. R. & E. M. Edmonds. (1978). Teacher vagueness and pupil participation in mathematics learning. *Journal for Research in Mathematics Education*, **9**, 228–232.

Smith, L. R. & M. L. Land. (1980). Student perception of teacher clarity in mathematics. *Journal for Research in Mathematics Education*, **11**, 137–146.

Solow, D. *How to read and write proofs*.

Steen, L. A. (Ed.) (1987). *Calculus for a new century: A pump, not a filter*. Washington, D. C.: Mathematical Association of America.

Steen, L. (1989). *Reshaping college mathematics: A project of the Committee on the Undergraduate Program in Mathematics*. Washington, DC: The Mathematical Association of America.

Stein, M. K., B. W.Grover & E. A. Silver. (1991). Changing instructional practice: A conceptual framework for capturing the details. In R. G. Underhill (Ed.), *Proceedings of the Thirteenth Annual Meeting, North American Chapter of the International Group for the Psychology of Mathematics Education*, **1** (pp. 36–42). Blacksburg, VA: VPI.

Sterling, J. & M. W. Gray. (1991). The effect of simulation software on students' attitudes and understanding in introductory statistics. *Journal of Computers in Mathematics and Science Teaching*, **10**, 51–56.

Stones, I., M. Beckmann & L. Stephens. (1983). Factors influencing attitudes toward mathematics in pre-calculus college students. *School Science and Mathematics*, **83**, 430–435.

Suydam, M. N. (1975). *Compilation of research on college mathematics education*. Columbus, OH: ERIC Information Analysis Center for Science, Mathematics and Environmental Education.

Suydam, M. (1989). Research on mathematics education reported in 1988. *Journal for Research in Mathematics Education*, **20**, 379–426.

Sweller, J. (1990). On the limited evidence for the effectiveness of teaching general problem–solving strategies. *Journal for Research in Mathematics Education*, **21**, 411–415.

Tall, D., & R. L. Schwarzenberger (1978). Conflicts in the learning of real numbers and limits. *Mathematics Teaching*, **83**, 44–49.

Tall, D. & S. Vinner. (1981). Concept image and concept definition in mathematics with particular reference to limits and continuity. *Educational Studies in Mathematics*, **12**, 151–169.

Tall, D. (1985). Using computer graphics programs as generic organizers for the concept image of differentiation. In L. Streefland (Ed.), *Proceedings of the Ninth International Conference for the Psychology of Mathematics Education*, **1**, (pp. 105–110). Utrecht, The Netherlands: State University of Utrecht.

Tall, D. (1987). Constructing the concept image of a tangent. In J.C. Bergeron, N. Herscovics, & C. Kieran, (Eds.), *Proceedings of the 11th International Conference of Psychology in Mathematics Eduation*, (pp. 69–75). Montreal, Canada.

The Task Force on Women, Minorities, and the Handicapped in Science and Technology. (1989). *Changing America: The new face of science and engineering. Final report*. Washington, DC.

Thomas, G.B. (1976). *Elements of calculus*. Reading, MA: Addison Wesley.

Thomas, G. E. (1986). Cultivating the interest of women and minorities in high school mathematics and science. *Science Education,* **70,** 31–43.

Thomas, W. E. & D. A. Grouws. (1984). Inducing cognitive growth in concrete–operational college students. *School Science and Mathematics,* **84,** 233–243.

Thompson, A. G. (1989). Learning to teach mathematical problem solving: Changes in teachers' conceptions and beliefs. In R. I. Charles & E. A. Silver (Eds.) (1989). *Research agenda for mathematics education: The teaching and assessing of mathematical problem solving* (pp. 232–243). Reston, VA: National Council of Teachers of Mathematics.

Thornton, M. C. & R. G. Fuller. (1981). How do college students solve proportional problems? *Journal of Research in Science Teaching,* **18,** 335–340.

Tinto, V. (1987). *Leaving college.* Chicago: University of Chicago Press.

Tobin, K. & M. Espinet. (1990). Teachers as peer coaches in high school mathematics. *School Science and Mathematics,* **90,** 232–244.

Treisman, P. U. (1985). *A study of the mathematics performance of Black students at the University of California, Berkeley,* Unpublished doctoral dissertation, Berkeley, CA.

Trelinski, G. (1983). Spontaneous mathematization of situations outside mathematics. *Educational Studies in Mathematics,* **14,** 275–284.

Triola, M. (1989).*Elementary statistics,* 4th ed. Redwood City, CA: Benjamin/Cummings.

Tucker, T. (Ed.) (1990). *Priming the calculus pump: Innovations and resources.* (Mathematical Association of America Notes No. 17). Washington, DC: Mathematical Association of America.

Turner, G. (December 1988) Increasing learning by decreasing math anxiety. *Humanistic Mathematics Network,* **3.**

Valverde, L. A. (1984). Underachievement and underrepresentation of Hispanics in mathematics and mathematics–related careers. *Journal for Research in Mathematics Education,* **15,** 123–133.

Vinner, S. (1983). Concept definition, concept image, and the notion of function. *International Journal of Mathematics Education in Science and Technology,* **14,** (3), 293–305.

Vinner, S. (1987). Continuous functions — images and reasoning in college students. In J. Bergeron (Ed.), *Proceedings of the 11th International Conference on the Psychology of Mathematics Education,* Montreal: University of Montreal.

Vinner, S. (1989). The avoidance of visual considerations by calculus students. *Focus: On Learning Problems in Mathematics,* **11,** 149–156.

Vinner, S. & T. Dreyfus. (1989). Images and definitions for the concept of function. *Journal for Research in Mathematics Education,* **20,** 356–366.

von Glasersfeld, E. (1984). An introduction to radical constructivism. In P. Watzlawick (Ed.) *The invented reality.* NY: Norton.

Wagner, S. & C. Kieran (Eds.) (1989). *Research agenda for mathematics education: Research issues in the learning and teaching of algebra.* Reston, VA: National Council of Teachers of Mathematics.

Wagner, W., S. L. Rachlin, & R. J. Jensen. (1984). *Algebra learning project: Final report.* Athens, GA: University of Georgia, Department of Mathematics.

Walter, R. L. (1973). *The effect of the knowledge of logic in proving mathematical theorems in the context of mathematical induction.* (Doctoral dissertation, The Florida State University, 1972). Dissertation Abstracts International, **33,** 2625A.

Watson, F. R. (Ed.) (1978). *Proof in mathematics.* Staffordshire: Institute of Education, University of Keele.

Wiener, H. S. (1986). Collaborative learning in the classroom. *College English,* **48** (1), 52–61.

Williams, B. R. (1972). *Critical Thinking ability as affected by a unit on symbolic logic.* (Doctoral dissertation, Arizona State University, 1971). Dissertation Abstracts International, **31,** 6434A.

Williams, S. R. (1991). Models of limits held by college calculus students. *Journal for Research in Mathematics Education,* **22** (3), 219–236.

Wittgenstein, L. (1983) *Remarks on the Foundations of Mathematics.* Cambridge, MA: MIT press,

Zorn, P. (1990). Algebraic, graphical, and numerical computing in elementary calculus: Report of a project at St. Olaf College. In F. Demana & J. G. Harvey (Eds.). *Proceedings of the Conference on Technology in Collegiate Mathematics.* Menlo Park, CA: Addison–Wesley Publishing Company.